SS

WEAPONS
and
FIGHTING
TACTICS
of the
WAFFEN-SS

SS
WEAPONS
and
FIGHTING
TACTICS
of the
WAFFEN-SS

Dr S. Hart & Dr R. Hart

British Library Cataloguing in Publication Data:
A catalogue record for this book is available
from the British Library

Copyright © 1999 Brown Packaging Books Ltd

ISBN 1-86227-060-0

First published in the UK in 1999 by
Spellmount Limited
The Old Rectory
Staplehurst
Kent TN12 0AZ

1 3 5 7 9 8 6 4 2

Editorial and design: Brown Packaging Books Ltd
Bradley's Close, 74-77 White Lion Street,
London N1 9PF

Editor: Philip Trewhitt
Design: Wilson Design Associates
Picture Research: Ken Botham

Printed in Italy

Picture credits:
Bundesarchiv: 2-3, 62, 64, 84, 102, 108, 112, 115, 116, 119, 122, 174
TRH Pictures: front cover, 11, 14, 17, 20, 28, 30, 38, 39, 43, 45, 47, 48,
55, 56, 58, 60-61, 68, 69, 71, 73, 75, 76, 78-79, 81, 86, 89, 90, 94-95,
98-99, 100, 106-107, 110, 125, 126, 133, 136, 138-139, 143, 144-145,
148, 150, 153, 154, 156, 158-159, 162, 167, 176, 178, 182, 186-187, back cover
TRH Pictures via Espadon: 6, 9, 12, 18, 23, 26, 32, 35, 61, 82, 93, 104,
128, 171, 188
Artwork credits:
All artworks Bob Garwood

Previous pages: Waffen-SS Panzer IVs on parade

Contents

Introduction

During World War II the Waffen-SS gained a reputation as being one of the modern world's most formidable, yet also brutal and evil, military élites. This work examines the myth and reality of the Waffen-SS: the battlefield triumphs and the atrocities, plus the empire-building ambitions of its head, Heinrich Himmler, that eroded its combat effectiveness.

The Schutzstaffel (SS) was the vanguard political force of Adolf Hitler's National Socialist (Nazi) Party that came to power in Germany in 1933. The Nazis attempted to create a new 'European Order' in which the Third Reich would endure for a thousand years. The SS comprised those individuals most committed to the Nazis' racist, violent and expansionist ideology. The Waffen-SS (literally 'armed SS') was the military arm of the SS that came into existence during March 1940. This organisation was created by an amalgamation of three existing SS military and paramilitary institutions: the *SS Leibstandarte Adolf Hitler* (the Führer's bodyguard regiment),

LEFT: *SS-Obergruppenführer Josef 'Sepp' Dietrich, commander of I SS Panzer Corps, inspects SS pistol training in the field at Kursk, Russia, summer 1943.*

the *SS-Verfügungs* (special readiness) Division and the Totenkopf (Death's Head) concentration camp guard units. These Waffen-SS formations were given names rather than numbers to proclaim their status as a military élite.

Two characteristics dominated the activities of the armed SS during World War II. First, there was the superb Waffen-SS fighting machine. Often its soldiers were brave, determined and skilled; they also frequently displayed excellent camaraderie and *esprit de corps*. This was due in part to the excellent equipment that the premier Waffen-SS divisions received during the latter half of the war, although less important formations had to make do with whatever weapons were available.

The second characteristic was the Waffen-SS as a brutal, fanatical, politicised force fully committed to the Nazis' ideological goals. The

Waffen-SS epitomised the terror that was at the heart of Nazism. Imbued with such views, many Waffen-SS soldiers arrogantly viewed themselves as being racially superior warrior-heroes destined to carve out the new Nazi European Order by annihilating the Reich's enemies. The Waffen-SS conceived life as a bitter struggle for racial and national survival, and hence embraced ruthlessness and iron will as the foundations of its institutional ethos. Consequently, Waffen-SS soldiers committed numerous atrocities against unarmed enemy prisoners and innocent civilians, in contravention of international law. The ideological indoctrination that Waffen-SS soldiers underwent created this combination of battlefield prowess and ruthlessness meted out to the enemy. As the vanguard of the Nazi movement, the Waffen-SS participated in Germany's expansionist wars and then spearheaded the maintenance of the internal security of the Nazi New Order in Europe.

THE MYTH OF THE WAFFEN-SS

Although the current popular fascination with the Waffen-SS partly reflects a widespread interest in élite forces, it is still one with disturbing aspects. The literary focus on images of handsome, blond-haired, blue-eyed and black-uniformed Aryan SS 'supermen' obscures the less palatable reality of the organisation. Too much literature has focussed on the impressive fighting record of the Waffen-SS but has down-played the organisation's darker side – its blatant racism and the hideous atrocities committed by its troops – as aberrant lapses in the heat of battle. But the military prowess of the Waffen-SS and its racist brutality were inextricably inter-linked: both were the product of the organisation's ideological motivation. Indeed, to examine the weaponry of aggression used by the Waffen-SS without emphasising the evil ends the organisation often put them to, would itself be a crime.

The popular perception of the Waffen-SS, therefore, is far removed from reality. For there was simply no such thing as a typical Waffen-SS soldier or formation. By the end of the war the Waffen-SS had evolved into a barely recognisable instrument compared to the fledgling organisation of 1940. By 1945 the armed SS had become a vast, racially heterogenous organisation, nearly a million strong. The popular view of the Waffen-SS as a Nazi Aryan élite pertains only to a few of its formations: the 'classic' divisions, like the *Leibstandarte, Das Reich, Totenkopf, Wiking* and *Hitlerjugend*. These constituted an élite within an élite that displayed an *esprit de corps* and fighting record second to none during World War II.

DILUTION OF RACIAL IDEALS

The Waffen-SS had a chequered history, though. During 1939–41, it struggled in the face of entrenched German Army opposition, firstly for the right to bear arms and then to receive the weapons and equipment necessary for it to become a military élite. Over the remainder of the war, the once wholly Aryan Waffen-SS gradually grew into a pan-national, multi-ethnic conglomeration. In 1941 the Waffen-SS opened its doors to 'Germanic' Scandinavians; by 1945 the exigencies of war found all sorts of, in Nazi eyes, racially inferior peoples serving with the armed SS, including Muslim Bosnians, Ukrainian and Russian Slavs, and Cossacks. Equally, some of the later SS 'divisions' raised in the last six months of the war were little more than a heterogenous mix of conscripts, remnants, stragglers and former prisoners of war.

The transformation, or rather degeneration, experienced by the Waffen-SS during 1939–45 was a result of a combination of two forces: the grandiose empire-building designs of the Reichsführer-SS – the former chicken farmer Heinrich Himmler – and the necessities of war, particularly the Germans' desperate need for manpower.

The wartime Waffen-SS thus developed in a number of distinct stages. Between 1939 and 1942, the Waffen-SS was a small, predominantly

motorised élite force. The premier Aryan SS formations – *Leibstandarte, Das Reich* and *Totenkopf* – were organised as motorised infantry divisions and equipped with rifles, machine guns, mortars, infantry guns, anti-tank weapons, artillery and armoured cars. It was these three divisions that established the military reputation of the Waffen-SS with distinguished service in appalling and brutal fighting conditions on the Eastern Front during 1941–42. The defensive resilience that these SS divisions displayed impressed Hitler and led him both to expand the Waffen-SS and to enhance the organisation's élite status.

However, by 1942 Himmler's desire to expand his SS empire had already weakened the coherence of the Waffen-SS. Himmler raised several new, less well-equipped formations that performed poorly in combat, but which spearheaded the brutal racial war the Nazis conducted behind the front in the East. The SS Brigade

ABOVE: *A Waffen-SS machine-gun crew, armed with a 7.92mm MG 34, on the Eastern Front in 1943. Note the SS-pattern camouflage smocks being worn.*

Nord, hastily raised from Totenkopf concentration camp guards, disgraced the Waffen-SS with its panicked rout by the Red Army at Salla, Finland, in July 1941. The SS Riding Brigade – the precursor of the 8th SS Cavalry Division *Florian Geyer* – again drawn from Totenkopf units, undertook a mass murder campaign to cleanse the captured Eastern territories which were to provide the Lebensraum (Living Space) necessary for the flourishing of the Aryan race. Waffen-SS troops also served with the genocidal Einsatzgruppen, the murder squads which executed 1.5 million Jews, Bolsheviks and other 'racial undesirables' in the East during 1941–43.

During 1942 the three premier Waffen-SS formations, plus the *Wiking* Division comprised of 'Germanic' volunteers, converted to mechanised infantry (panzergrenadier) divisions. During this process, the Waffen-SS received its first large-scale allocation of tanks and assault guns, and hence finally gained a significant offensive capability. During this period other new weapons joined the SS inventory, including Marder self-propelled anti-tank guns, together with Wespe and Hummel self-propelled howitzers. The three premier SS divisions returned to the Eastern Front during the spring of 1943 to spearhead the successful German counter-strikes that recaptured Kharkov. This battlefield achievement further enhanced the military reputation of the Waffen-SS, but this accomplishment was also sullied by more atrocities.

WAFFEN-SS PANZER DIVISIONS

During 1943, the Waffen-SS experienced renewed expansion. The *Leibstandarte, Das Reich, Totenkopf* and *Wiking* Divisions became full panzer formations and began to receive Germany's latest Panther and Tiger tanks. That year the Waffen-SS raised three additional armoured divisions: *Hohenstaufen, Frundsberg* and *Hitlerjugend*, as well as a new 'Germanic' SS Panzergrenadier Division *Nordland*. These formations ultimately developed a fighting prowess akin to that exhibited by the classic SS formations. Himmler's expansionist ambitions, however, further diminished the organisational unity of the Waffen-SS as the number of its formations burgeoned during 1943–44. During the autumn of 1943, increasing diversity and growing terminological confusion compelled Himmler to number SS divisions in strict order of their formation.

Himmler's expansionism led to the raising of new SS formations during 1943–44, which tapped the ethnic Germanic populations of central Europe instead of competing directly with the rest of the armed forces for scarce German manpower. These formations included the 16th, 17th and 18th SS Panzergrenadier Divisions *Reichsführer-SS, Götz von Berlichingen* and *Horst Wessel*. Although these divisions per-

formed reasonably well in combat, they never acquired the same battlefield resilience displayed by the premier formations. Equal to these divisions, in terms of combat effectiveness, were the best of the foreign volunteer formations, which were raised in ever increasing numbers after 1942. Such formations included the Flemish troops of the 27th SS Volunteer Panzergrenadier Division *Langemarck*, and the Balts of the 15th SS Volunteer Grenadier Division (Latvian No 1).

Below these forces in terms of combat performance were the racially mixed divisions formed during 1943–45, such as the 21st and 24th SS Volunteer Mountain Divisions *Skanderbeg* (Albanian No 1) and *Karstjäger*, respectively. Many of these formations possessed limited combat value and were useful only for brutal anti-partisan sweeps in the rear areas. Finally, there was an underclass of formations with terrible records for brutality, even by Waffen-SS standards. This category included depraved outfits like the *Kaminski* and *Dirlewanger* Brigades.

NOT JUST A MILITARY FORCE

Since 1945, apologists for the Waffen-SS have examined the organisation simply as a battlefield military élite. Such a narrow perspective obscures the undeniable truth of the organisation's dual politico-military character, and its consequent full participation in the Nazi Holocaust. Waffen-SS troops spearheaded the bloody anti-partisan operations undertaken throughout Nazi-occupied Europe, participated in the genocide perpetrated by the Einsatzgruppen, and if invalided often served as guards at the heinous concentration and extermination camps. The armed SS was thus never just a military force, but the political vanguard of the Nazis' genocidal goals in Europe.

Given this reality, it may perhaps seem surprising that during 1939–42, the Waffen-SS often fought with second-rate and captured foreign weapons. This situation was largely due to the continuing resistance of the German Army to SS procurement demands. Thus it was only in 1942, after it had proven its resilience in bitter defensive fighting in the East during the winter of 1941–42, that the Waffen-SS finally began to receive quality weaponry in quantity. Even then it was not until 1943, when the Waffen-SS had unequivocally established itself as a military élite, that it acquired Germany's latest and most effective equipment. During 1944-45, the premier SS divisions received priority over army formations for the allocation of Germany's latest weapons.

During 1944–45 the Waffen-SS took on a decidedly defensive posture, and a host of new weapons entered service, but never in sufficient quantity to make a difference. Cheaper, quicker-to-build assault guns and tank destroyers progressively replaced tanks in SS panzer and panzergrenadier divisions; heavy mortars and light field guns replaced infantry guns and field howitzers; and anti-tank rocket launchers replaced anti-tank guns. But in the final analysis, no amount of either technological innovation or battlefield fanaticism could ultimately offset the vastly unequal odds that Germany faced in 1943–45. Such overwhelming inferiority was the product of the strategic arrogance of the Nazi leadership, which was shared by the self-proclaimed Aryan supermen of the Waffen-SS. But, during the spring of 1945, the Allies conquered Germany and then brought to justice at Nuremberg the surviving Nazi and Waffen-SS leaders who had committed war crimes during World War II.

BELOW: *A 7.5cm leIG 18 light infantry gun of the* SS-Verfügungs *Division in action in France in May 1940. Note the large tyres that allowed it to be towed.*

CHAPTER 1

Small Arms

The soldiers of the Waffen-SS developed a unique bond with their personal weapons. Special swords, daggers and pistols enhanced the mystique that surrounded the Waffen-SS as an élite. But it was modern submachine guns and assault rifles in the hands of SS soldiers that buttressed their ideologically based fanaticism and made them the 'Supermen' of Nazi ideology.

Swords and daggers played a central role in the pomp and pageantry that accompanied Nazi martial spectacle. To help distinguish the Waffen-SS as an élite, its members carried their own distinct daggers and swords. The SS dagger was awarded to all personnel on the completion of three years' service and came in two basic forms: the 1933 and 1936 editions. The earlier model was of superior quality: it was inscribed with the individual's SS number, and carried on its obverse the SS motto 'Loyalty is my Honour' inscribed in gothic script. The 1936 version was less ornate and omitted the individual's SS number. These changes allowed the mass production

LEFT: *An SS grenadier in a dugout on the Eastern Front. He is armed with a 7.92mm Karabiner 98K, the standard Waffen-SS service rifle of World War II.*

which became necessary after the late 1939 expansion of the two SS paramilitary organisations – the Verfügungstruppe (special readiness troops) and the Totenkopfverbände (Death's Head concentration camp guards) – which merged in early 1940 to form the Waffen-SS.

Both daggers had a blackened wood hilt with a steel cross-guard and pommel and resembled a German hunting knife in design. A small silver eagle clutching a swastika in a wreath was inset into the centre of the obverse, while the SS runes were impressed in a small circle at the base next to the pommel on the same side.

Small numbers of special silver-plated SS honour daggers were also issued for meritorious service. These were of both the 1933 and 1936 pattern, and were intricately stamped with acorns and oak leaves. They were issued in small numbers, and even rarer was the SS dedication dagger presented for personal services rendered to the head of the SS, Reichsführer-SS Heinrich Himmler. The dedication dagger bore Himmler's personal signature and inscription on the reverse of the blade. He first issued the dagger in 1934 to SS officers who had participated in the January 1934 'Night of the Long Knives', when Hitler used the SS to purge the leadership of the Sturm-abteilung (SA), its paramilitary rival. The irony of this issuance was that the idea for a special SS dedication dagger originated from Ernst Röhm, the head of the SA, whom the SS murdered during the 'Night of the Long Knives'.

BELOW: *The* SS Leibstandarte Adolf Hitler *on parade at the Berlin-Tempelhof airport, March 1935. Note the straight sword carried by the NCO on the left.*

Waffen-SS personnel also received a range of special swords. Indeed, while duelling was officially banned in the Third Reich – Hitler considered it an anachronistic product of the Prussian martial élitism he so detested – fencing and swordsmanship were skills prized by the Waffen-SS. Indeed, SS-Gruppenführer (Lieutenant-General) Reinhard Heydrich, the head of the SD (SS Security Service), was an Olympic fencer. Dozens of different swords and honour swords were commissioned for the SS. There were three basic designs, however: the non-commissioned officer's sword; the officer candidate's sword; and the officer's straight sword that replaced the earlier officer's curved sabre in 1936. All these swords possessed the common D-shaped guard and a black hilt. The NCO's sword had the SS runes inscribed on an otherwise plain pommel,

Mauser C96 Pistole

German Designation: Mauser C96 Pistole
Weapon Type: Pistol
Calibre: 7.63mm (0.31in)
Magazine Capacity: 10-20 rounds
Length: 64.7cm (2ft 4in)
Weight: 1.78kg (3.9lb)
Muzzle Velocity: 480m/s (1574ft/s)

while the officer's sword had the runes inset in a large circle on the centre of the obverse of the hilt. The officer's and officer candidate's swords also had their pommels decorated. A few swords had inscriptions engraved on them to commemorate some special achievement, such as the German hand-grenade throwing record which was held by an SS trooper.

The earlier officer's sabre was a more attractive sword and was prized by SS officers. The pommel formed an eye-catching lion's head and the back-strap of the hilt was decorated with silver oak leaves inset with SS runes. The cross-guard also incorporated the Nazi eagle and swastika. Small numbers of honour sabres were also issued prior to 1936. These were distinguishable from the regular sabre by the addition of the SS motto inscribed on the obverse of the blade. One other distinctive sword that saw limited service in the SS was the German Army's 'Prinz Eugen' sword which was worn, by special dispensation, exclusively by officers and NCOs of the 7th SS Volunteer Mountain Division *Prinz Eugen*.

PISTOLS IN WAFFEN-SS SERVICE
Waffen-SS pistols were almost entirely of the automatic or self-loading variety, the revolver having gone out of German military fashion in the early twentieth century. German pistols were generally sophisticated and of extremely high

Parabellum 1908

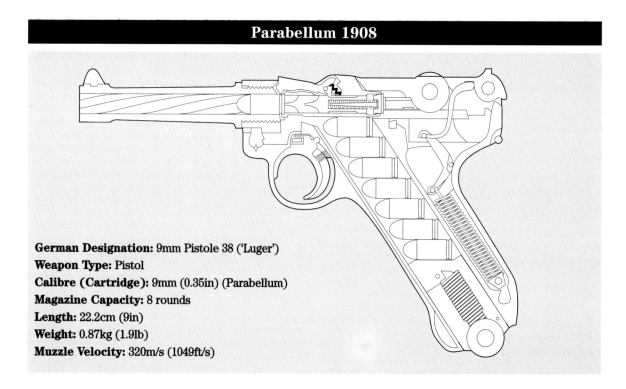

German Designation: 9mm Pistole 38 ('Luger')
Weapon Type: Pistol
Calibre (Cartridge): 9mm (0.35in) (Parabellum)
Magazine Capacity: 8 rounds
Length: 22.2cm (9in)
Weight: 0.87kg (1.9lb)
Muzzle Velocity: 320m/s (1049ft/s)

quality, durability and reliability. However, German industry could never meet the demand for pistols, and so the Waffen-SS had to supplement its standard service pieces with a wide range of captured ones. Though a pistol became a prestige symbol in the Waffen-SS, in reality pistols possessed little combat value because they had only a short range and limited accuracy and were thus really only useful for close self-defence. But as a status symbol the pistol performed an important role in enhancing the mystique of the Waffen-SS as a military élite. Only amid the brutal fighting on the Eastern Front, where the Nazi racial-ideological war created a bitter cycle of atrocity and reprisal, did the pistol perform an important combat role: a 'friendly' bullet was an ending that many SS troopers reluctantly chose over the unpleasant prospect of being a Soviet prisoner of war!

The two pistols which saw most common service with SS personnel were the standard armed forces issue 9mm Pistole 08 and 9mm Walther Pistole 38. The former, widely known outside Germany as the Luger, has become associated in the popular consciousness with the Waffen-SS. Despite its popularity with collectors and Hollywood, the Luger was a far-from-ideal service pistol. Designed at the turn of the century, it entered German service in 1908. It was a recoil-operated, single-shot automatic with a toggle lock and manual safety switch. The sliding box magazine held eight rounds, though a few elongated-barrel Lugers carried a special 32-round magazine that turned the pistol into a machine carbine. The Luger had a maximum effective range of 50m (55yds), but its toggle breech mechanism was open to the elements and easily became fouled. It thus required careful handling and frequent cleaning. Also, the high quality of the weapon made mass production time-consuming and expensive.

Thus in 1938 the German military introduced a replacement pistol, the Walther Pistole 38. It included a double-action trigger mechanism and was specifically designed for mass production. In service it proved an excellent weapon, so much so that the West German Army placed it back into production in 1957. Nevertheless, the Pistole 38 never entirely supplanted the Pistole 08 and both remained in SS service until the end of the war. Despite production of more than one million Pistole 38s during the war, the pistol was never a particularly common weapon, and the numbers available were well below operational requirements. The Walther P38 proved such a good side-arm that SS-Oberführer (Colonel) Gürttner, the head of the SS Procurement Office, fought a long, bitter but ultimately unsuccessful campaign with the Army Weapons Department

to get the entire production of the pistol diverted to the Waffen-SS. Though not as elegant and graceful as the Luger, the stubbier Walther was easier to strip and maintain and its performance was essentially identical.

WALTHER PISTOLS

Because demands for the Pistole 38 rapidly outstripped supply, the Waffen-SS took into service numbers of the earlier Walther PP pistol. Indeed, many of the officers and NCOs of the *SS-Polizei* Division raised in 1939–40 from the Order Police

BELOW: *A member of the* Leibstandarte *Division armed with a 9mm Walther Pistole 38 pistol at Kharkov during the spring of 1943.*

ABOVE: *Two Waffen-SS soldiers on guard duty in Hungary during the winter of 1944–45. The soldier on the left carries a Karabiner 98K rifle.*

carried this weapon. The SS also adopted the shorter-barrelled Walther PPK, which officers favoured because it could be easily concealed. The PPK became the preferred weapon of the personnel of affiliate branches of the SS empire – the Algemeine-SS, the SD and the Gestapo. Walther first introduced the PP in 1929, and it continued to be manufactured throughout the war in 6.35mm, 7.65mm and 9mm versions. Both the PP and PPK were of advanced design for their time, with blow-back action, external hammers, double-action triggers and safety guards. Though very easy to strip in the field, they suffered from the drawback of only having a range of 25m (27yds).

Another pistol that saw less frequent use among the Waffen-SS was the Mauser C96, the so-called 'Broomhandle' Mauser. First produced in 1896, it became one of the most famous pistols of all time. Its bulk, relative complexity and decreasing numbers ensured that it usually served only with rear-area SS forces, but it proved valuable because of its impressive fire-power. Despite the fact that it was superseded by modern, purpose-built submachine guns, it remained a sought-after weapon, not for its effectiveness but because its distinctiveness accentuated the élite aura of the SS. The pistol was produced in both 7.63mm and 9mm calibres, as well as a specialised rapid-fire automatic version that featured a 20-round magazine and a wooden shoulder stock. Despite the fact that it was virtually uncontrollable when firing full-automatic, it remained popular with SS security forces on anti-partisan missions, where questions of morale and firepower were often more important than accuracy.

Because German-produced pistols were frequently in short supply, the Waffen-SS often had to equip itself with captured foreign weapons. Czech pistols saw widespread service in the SS during 1939–41. The most commonly carried Czech pieces included the 7.65mm Pistole 27(t) and the 9mm Pistole 39(t). The latter was the main Czech service pistol, the 9mm CZ 38, but it had a rather clumsy double-action design and a very heavy trigger-pull which hampered accuracy. Its general unpopularity among the German armed forces ensured that most of the pistols found their way into SS hands, where they were commonplace among the *SS-Polizei* Division and the SS Brigade *Nord*. A much more modern service pistol that joined the German small arms arsenal late in 1939 was the Polish 9mm Radom vis wz 35.

In 1940 another of the world's leading small arms manufacturers, the Belgian Fabrique Nationale d'Armes de Guerre at Herstal, Liege, fell into German hands. It produced a variety of excellent self-loading pistols, including the 1900, 1903, 1910 and 1922 FN models, which all saw limited SS service. One of its most renowned

side-arms was the Browning Grande Puissance pistol. Designed in 1925, it entered Belgian service in 1935 and the Germans kept it in production, under the designation 9mm Pistole 640(b), until 1944. The pistol remains in production today and has become one of the most widely utilised pistols of all time. The Browning possessed similar performance to the German Luger and Walther, the only major difference being a larger 13-round clip which was made possible by broadening the handle, so that the bullets could be arranged in two staggered rows. Another Browning pistol used by the Waffen-SS was the smaller 9mm Model 1910. With a short 89mm (3.6 in) barrel and seven-round clip, it was comparable to the Walther PPK in terms of characteristics and performance.

BALKAN PISTOLS

Another fertile source of pistols for the SS was the Balkans. The three Hungarian SS divisions raised late in the war received issues of the standard Hungarian service pistol, the 7.765mm Pisztoly 37M, designated the 7.765mm Pistole 37(u) in German service. The weapon continued in production in Budapest after the Soviet encirclement of the Hungarian capital, until the

Karabiner 98b

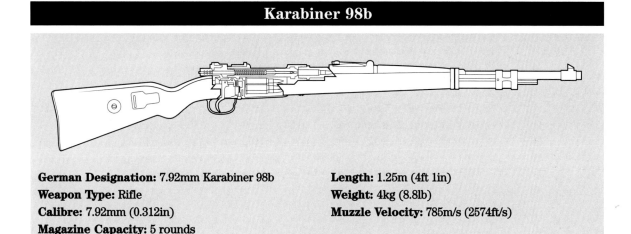

German Designation: 7.92mm Karabiner 98b
Weapon Type: Rifle
Calibre: 7.92mm (0.312in)
Magazine Capacity: 5 rounds

Length: 1.25m (4ft 1in)
Weight: 4kg (8.8lb)
Muzzle Velocity: 785m/s (2574ft/s)

LEFT: *Russia, 1944, and an excellent view of a Sturmgewehr 44 assault rifle. This fine weapon was concentrated in élite Waffen-SS counterattack units.*

city fell to the Red Army in February 1945. Newly manufactured pistols were issued to Budapest's defenders during the winter of 1944–45 to supplement their dwindling weapon stocks. From the large Waffen-SS contingent of the doomed garrison – the 8th SS Cavalry Division *Florian Geyer* and the 22nd SS Volunteer Cavalry Division *Maria Theresia* – just 170 SS troops managed to fight their way successfully through to German lines! No doubt, some of the encircled SS troops followed the example set by the wounded commander of the *Florian Geyer* Division, SS-Oberführer (Colonel) Joachim Rumohr, when on 12 February 1945 he chose suicide with his Pisztoly 37M pistol rather than surrender to the Soviets.

PISTOLS FROM ALLIES

A side-arm which was popular among Austrian SS officers of the *Der Führer* Regiment of the 2nd SS Panzer Division *Das Reich* was the elegant Hungarian 7.65mm Frommer pistol, which had an unusual long recoil mechanism, the barrel traversing fully to the rear of the pistol. The advantage of such a recoil was minimal kick on firing and hence enhanced accuracy. Norwegian SS volunteers of the 5th SS Panzer Division *Wiking* made extensive use of the 11.43mm Model 1914 Norwegian service pistol, a license-built copy of the famous American Colt M1911 automatic. Its large calibre made the weapon greatly prized among SS officers, despite the problem of acquiring suitable ammunition. The pistol was a recoil-operated automatic with a seven-round magazine. Despite great stopping power, the gun was heavy and the recoil harsh, making it a difficult weapon to aim and fire.

Friendly foreign countries also supplied pistols to the SS, such as the Spanish Astra Mod. 400 pistol, the 9mm Pistole Astra and the 9mm Mod. 600 Astra. Small numbers of these side-arms saw service with SS units, mainly in Western Europe.

With armaments production increasingly diverted to more important weapons, supplies of pistols remained far below operational requirements. At the start of the war in September 1939, for example, the four SS motorised infantry regiments were authorised some 3812 pistols, against a personnel strength of 12,670 officers and other ranks. Once SS units joined combat, attrition inevitably further decreased the number of pistols available.

BOLT-ACTION RIFLES

The most common weapon used by SS soldiers during World War II was the bolt-action rifle. While frontline SS troops invariably carried the basic German armed forces rifle, the Gewehr 98, second-line and foreign volunteer SS units used a bewildering array of captured rifles. This weapon and its shorter cousin designed originally for cavalry use, the carbine, had been standard infantry equipment since the nineteenth century.

The Gewehr 98 was based around one of the most efficient and reliable bolt-actions ever made – the Mauser system – and proved to be one of the most influential rifles of all time. The Gew 98, an improved version of a rifle first developed in 1888, remained in production until 1945 and was widely copied by other nations. It was a single-shot, bolt-action rifle of 7.92mm calibre that carried a five-round magazine. The ammunition was inserted in clips into the magazine from above, which allowed half empty magazines to be 'topped up', a useful attribute. In contravention of the armaments limitations imposed by the 1919 Treaty of Versailles, the inter-war German military hid hundreds of thousands of Gew 98 rifles for future use. During the 1920s, the Gew 98 underwent a series of minor modifications that were the result of lessons learned in World War I. New ammunition and sights produced the 7.92mm Karabiner (Kar) 98b carbine, a modification of the original

Fallschirmjägergewehr 42

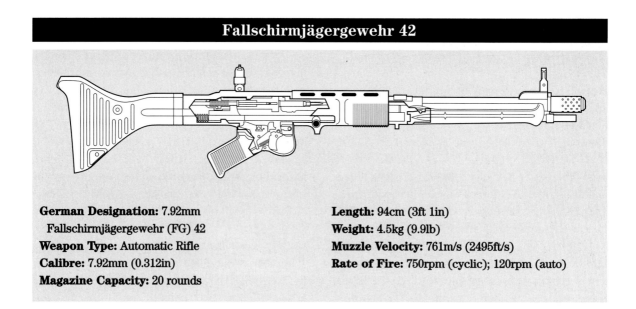

German Designation: 7.92mm
  Fallschirmjägergewehr (FG) 42
Weapon Type: Automatic Rifle
Calibre: 7.92mm (0.312in)
Magazine Capacity: 20 rounds

Length: 94cm (3ft 1in)
Weight: 4.5kg (9.9lb)
Muzzle Velocity: 761m/s (2495ft/s)
Rate of Fire: 750rpm (cyclic); 120rpm (auto)

Karabiner 98, the shortened cavalry carbine variant of the Gew 98 introduced in 1904. The Kar 98b went into mass production in the mid-1930s, and it was manufactured in such numbers that it became the most numerous German rifle of World War II, seeing widespread service until 1945. The shortened barrel of the Kar 98b was retained in the Czech-built version of the carbine that went into production in 1933. After Germany's occupation of Czechoslovakia in March 1939, the gun was pressed into German service as the Kar 33/40 and kept in production. Indeed, the carbine saw particularly wide use among Waffen-SS units during the early war years: it served with the 6th SS Mountain Division *Nord* in the Arctic Tundra of Finland and with the 7th SS Volunteer Mountain Division *Prinz Eugen* in the Balkans. During 1943–44, it served with the 13th, 21st, and 24th SS Volunteer Mountain Divisions *Handschar*, *Skanderbeg* and *Karstjäger* respectively, because its shortened barrel was more convenient for use on skis in mountainous terrain.

With the repudiation of the Treaty of Versailles in 1935 a new carbine, the 7.92mm

Karabiner 98k, entered production, and it became the most common German rifle of the later war years. However, due to increasing disruption of production by Allied air attack, the Kar 98k never supplanted either the Gew 98 or the Kar 98b, and all three rifles continued to serve with SS troops until the end of the war.

Rifles were numerically the most common weapon in the Waffen-SS arsenal. In September 1939, the four SS motorised infantry regiments were authorised some 9665 rifles against a ration strength of 10,867 enlisted personnel. As the Waffen-SS underwent a rapid expansion during the war, the number of rifles in its inventory skyrocketed, so that by the time the organisation peaked in size during the spring of 1945 at close to a million personnel, the Waffen-SS theoretically possessed hundreds of thousands of rifles. Since German production could never keep pace with demand from the front, the Waffen-SS often pressed into service a wide range of foreign rifles. So many foreign rifles saw service with the SS that it is impossible to describe them all here, and the following discusses only those that saw widespread and sustained use.

The Soviet Tokarev SVT 40 gas-operated, semi-automatic, self-loading rifle had a rate of fire of 25 rounds per minute, and was widely sought after by the Waffen-SS. Captured in large numbers along with plenty of ammunition during the opening stages of Operation 'Barbarossa' in 1941, the Tokarev became the first truly self-loading rifle in German service. It gave valuable service while German industry examined the weapon and developed its own range of self-loading rifles. To compensate for the gun's powerful recoil a tiny muzzle brake was fitted, but it had only marginal impact on reducing the rifle's fierce kick. Nevertheless, the Tokarev effectively combined the range of the rifle with the rate of fire of a submachine gun, though rapid fire could not be sustained for long. The gun weighed 3.9kg (8.6lb), took a 10-round magazine and had a maxi-

ABOVE: *A late-war study of Waffen-SS soldiers. The rear man carries a Sturmgewehr 44 assault rifle. Note the Flammenwerfer 41 flamethrower.*

mum range of 1500m (1640yds). The German experience with the Tokarev ultimately resulted in the emergence of an entirely new range of German self-loading rifles in the middle of the war.

Having discussed rifles, it is appropriate to say a few words about SS bayonets. Distinguishing individual SS bayonets is difficult, however, because SS troops used the three standard army issue bayonet – the model 84/98, 98/04, and 98/14 knife bayonets – all of which had been designed before World War I for the Gew 98 rifle and the Kar 98 carbine. The 84/98 bayonet was 386mm (1ft 3in) long with a straight 252mm (10in) blade that had a runnel on the face to facilitate extraction.

Contrary to popular opinion, hand-to-hand combat was actually a skill much neglected in early Waffen-SS training, as it did not accord with the fast Blitzkrieg warfare techniques SS recruits practised. The Waffen-SS therefore initially suffered a shock when its units encountered swarms of Soviet infantry during the summer of 1941 that made suicidal mass charges to engage the Germans in brutal hand-to-hand combat with bayonets, knives and entrenching tools. As a consequence, the Waffen-SS added close-combat instruction to its training curriculum and created the Gold Close Clasp, the highest infantry decoration short of the Knight's Cross, of which only 403 were ever awarded. Among those to receive the award was SS-Untersturmführer (2nd Lieutenant) Adolf Pechl of the 12th Company, 4th SS Panzergrenadier Regiment, 2nd SS Panzer Division *Das Reich*. He received this medal for surviving 50 days of close-combat on the Eastern Front despite being wounded. Indeed, he gained a reputation as a rare soldier who preferred the clash of steel on steel. Pechl went on to receive the Knight's Cross

in July 1944, again for close-quarter bayonet actions during bitter defensive fighting in Normandy's Bocage hedgerow terrain north of St. Lo. Surprisingly, despite his penchant for hand-to-hand combat, he survived the war!

Automatic, or more accurately self-loading, rifles first made a very limited appearance during World War I, but throughout the inter-war period the automatic rifle remained little more than a drawing-board design. While submachine guns could provide rapid fire at short ranges, what infantry needed was an automatic self-loading rifle that could provide rapid mass fire in the 250–750m (273–820yds) range where most infantry clashes occurred.

THE GEWEHR 41(W)

Mauser dusted off its experiments with self-loading rifles from World War I and married them to the gas-operated system used in the Soviet Tokarev rifle. In this design, the expanding gas created by firing a cartridge pushed back a piston enclosed in the barrel sleeve which in turn pushed a connecting rod which automatically chambered

Maschinenkarabiner 42

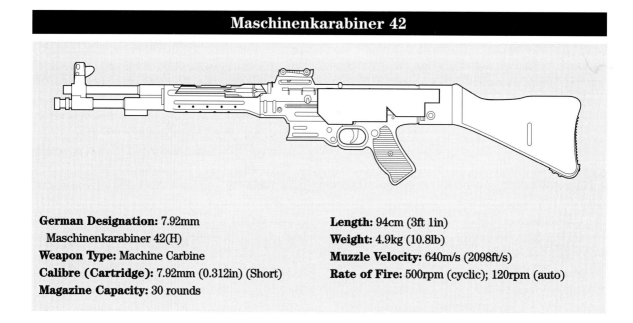

German Designation: 7.92mm Maschinenkarabiner 42(H)
Weapon Type: Machine Carbine
Calibre (Cartridge): 7.92mm (0.312in) (Short)
Magazine Capacity: 30 rounds

Length: 94cm (3ft 1in)
Weight: 4.9kg (10.8lb)
Muzzle Velocity: 640m/s (2098ft/s)
Rate of Fire: 500rpm (cyclic); 120rpm (auto)

a new cartridge. Walther joined the competition and its similar weapon, designated the 7.92mm Gewehr 41(W), proved marginally superior to the Mauser Gew 41(M). But the resulting self-loading rifle was clumsy and heavy and saw only limited production. The Gew 41 weighed 5kg (11.1lb) and took a 10-round box magazine which fired standard German 7.92mm ammunition out to a maximum effective range of 800m (880yds). The Gew 41 did not prove popular with troops in the field and their criticisms led to the lighter Gewehr 43. This weighed only 4.4kg (9.7lb) and had a shorter barrel. Making greater use of stamped and cast components, the Gew 43 was also easier and cheaper to produce.

At this point German experiments with self-loading rifles might have ended in failure, since Hitler opposed development of new weapons that fired pistol-calibre ammunition. Nevertheless, Hugo Schmeisser went on to develop Germany's first genuine self-loading carbine. He combined the new shorter, less powerful cartridge – the 7.92mm kurz patrone – with a new hybrid weapon, designated by the Germans as a machine-carbine. This weapon combined the full-automatic fire capability of the submachine gun, but with greater range.

THE MP 44 – THE SOLDIER'S WEAPON

The result was a hybrid automatic rifle that combined features of submachine gun, rifle and machine gun, designated the Maschinenkarabiner 42 (H). It was issued in small quantities to the SS Divisions *Leibstandarte*, *Das Reich* and *Totenkopf* in late 1942. An improved variant, designated the Maschinenpistole (MP) 43, entered service in late 1943. In the field it proved to be exactly what the troops had clamoured for, and in 1944 the MP 43 went into mass production, now redesignated (but virtually unmodified) MP 44. Finally, in the autumn of 1944, Hitler ordered the weapon's designation changed to the more appropriate Sturmgewehr 44 (assault rifle). The MP 43 was the

first of a genre of weapons that are known today as assault rifles. Although the Waffen-SS received first priority in the allocation of the Sturmgewehr 44, it remained a relatively uncommon weapon and only began to reach SS troops in any numbers in late 1944. The new 1945 organisation for SS grenadier divisions envisaged two platoons in each battalion to be equipped with the Sturmgewehr 44, though these goals proved impossible to realise. Instead, commanders typically grouped those MP 44 rifles that did reach their troops and issued them to élite assault platoons and companies, which were intended as formation counterattack reserves. SS assault units equipped with the Sturmgewehr 44 possessed far greater firepower than conventionally armed infantry. Manned by hand-picked personnel, typically the most decorated, experienced and hardened Eastern Front veterans, such units were held back to counterattack any enemy penetration of the German defences.

FIREPOWER DEFEATS MASS

One such MP 44-equipped 'fire brigade' unit was the 2nd Battalion, 26th SS Panzergrenadier Regiment, 12th SS Panzer Division *Hitlerjugend*. In early March 1945, the division launched a series of counterattacks to improve its positions in the Gran bridgehead on the southern sector of the Eastern Front in Hungary. Naturally, this battalion and its assault rifles spearheaded the attack on 7 March that stormed the town of Odon-Puszta in less than five hours, capturing a troop of heavy Soviet anti-tank guns in the process. Pressing on in winter snow, the battalion captured the important local communications centre of Igar, where it beat off repeated Soviet infantry attacks aimed at regaining the settlement. On 11 March, the battalion was ordered to storm the nearby Sio Canal and establish a bridgehead on the far side. In the face of determined resistance, it was not until nightfall than the battalion gained a foothold on the far

ABOVE: *A Leibstandarte reconnaissance troop in Poland, September 1939. The man in the foreground has a Bergmann MP 28 II submachine gun.*

bank of the canal, its assault rifles and Panzerfausts having finally smashed the enemy defences. Throughout the night, however, the battalion repulsed a series of Soviet counterattacks, and once again it was the volume of defensive fire generated by the battalion's MP 44 assault rifles that saved the situation.

Despite further Soviet counterattacks the battalion expanded its bridgehead to encompass the high ground of the nearby Hill 220, the loss of which goaded the enemy into their first major combined-arms assault. By the end of the day the battalion was down to 481 men and equipment losses had been even greater: the unit being

reduced to 192 rifles, 76 pistols, 22 submachine guns, 33 MP 44 assault rifles and 61 Panzerfaust anti-tank launchers. On 14 March, the battalion beat off another Soviet regimental attack but the following day a Soviet breakthrough at Stuhlweissenburg threatened the rearward communications of the SS division and it reluctantly had to evacuate the bridgehead to free up troops for a major counterattack. The next day the undefeated battalion retired in good order behind the canal, undisturbed by an enemy who had spent his strength trying to dislodge it.

THE FG 42 IN ACTION

An even more remarkable assault rifle was the Fallschirmgewehr 42 (FG 42) paratrooper weapon that saw service with the 500th and 600th SS Paratroop Battalions. This weapon was

Maschinenpistole 28

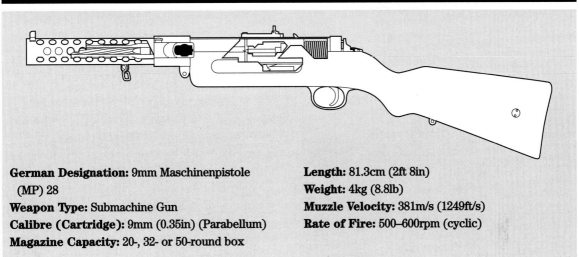

German Designation: 9mm Maschinenpistole
(MP) 28
Weapon Type: Submachine Gun
Calibre (Cartridge): 9mm (0.35in) (Parabellum)
Magazine Capacity: 20-, 32- or 50-round box

Length: 81.3cm (2ft 8in)
Weight: 4kg (8.8lb)
Muzzle Velocity: 381m/s (1249ft/s)
Rate of Fire: 500–600rpm (cyclic)

designed to meet the specific needs of German paratroopers for a light rifle, easily carried and stowed, that generated greater firepower than a conventional rifle. It was a weapon of light-weight construction that weighed just 4.5kg (10lb) and featured a side-mounted 20-round box magazine. It also had bipod legs attached to the barrel which allowed it to double as an effective light machine gun, in which guise it could fire an impressive 750-800 rounds per minute with an effective range of 1200m (1315yds). The weapon was complex and expensive to manufacture and Germany only produced 7000, but it was a great-ly prized. Skorzeny's SS commandos used the gun during their daring rescue of Benito Mussolini from imprisonment on Gran Sasso mountain after Italy's September 1943 capitula-tion. The 500th SS Parachute Battalion also used the FG 42 in its only wartime parachute drop on Josef Broz Tito's mountain cave headquarters at Drvar in Yugoslavia on 25 May 1944. The objec-tive of Operation 'Knight's Move' was to kill the leader of the Yugoslav Communist National Liberation Forces and thus emasculate his increasingly aggressive guerilla organisation.

Though Tito narrowly escaped with his life, the impressive firepower of the FG 42 allowed the greatly outnumbered SS paratroopers to hold off much larger partisan forces until a German ground relief column reached them.

SUBMACHINE GUNS

The Waffen-SS used submachine guns extensive-ly during World War II. A product of the trench warfare of the 1914–1918 war, the submachine gun evolved as a light, portable, one-man weapon to deliver a high rate of automatic fire, albeit over a short range. The first submachine gun was designed by a German weapons manu-facturer, Hugo Schmeisser, and designated the Maschinenpistole 18 Bergmann. The gun utilised a 32-round drum magazine, a rifle-style wooden stock and a perforated barrel sleeve to provide air-cooling, since the barrel became extremely hot with sustained fire.

During the inter-war period, a modified ver-sion of the Bergmann, designated the MP 28, entered German service. In the late 1930s a suc-cessor submachine gun, the MP 38, entered ser-vice. Produced by Erma-Werke, the weapon was

LEFT: *A Waffen-SS grenadier on the Eastern Front in the spring of 1942 wearing a fur-lined sheepskin overcoat and armed with an MP 38 submachine gun.*

(10.4lb), possessed an effective range of 200m (220yds), and was capable of automatic fire of up to 500 rounds per minute. Indeed, in combat the MP 40 proved to be one of the finest weapons of World War II. It was initially issued to platoon and section leaders, tank crews, and paratroopers, but its use ultimately became much more widespread, many SS grenadiers 'unofficially' acquiring the weapon. The separate SS procurement channels were ultimately able to secure a disproportionate percentage of the gun's production and thus the MP 40 saw wide service with the élite SS divisions.

CAPTURED SOVIET SUBMACHINE GUNS

Soviet submachine guns were widely used by the Waffen-SS personnel on the Eastern Front. The oldest of these was the 7.62mm PPD M1934/38/40. This weapon was based upon both the Bergmann MP 28 and the Suomi M1931, and resembled the former in appearance, though the quality of workmanship was much lower. The gun could carry either a 25-round box or a 71-round drum magazine, and it was capable of accepting a 7.63mm Mauser round. Another Soviet submachine gun that fell into SS hands was the PPSh 41, an even cruder and simpler version of the PPD that was rushed into mass production in late 1941. Produced in millions, it equipped the infantry of select Soviet shock regiments. It saw such widespread service, and was captured in such large numbers, that it ultimately became the second most numerous submachine gun in German service. One of the most remarkable of all Soviet weapons was the 7.62mm PPS 42 submachine gun, which was designed, like the British Sten, as an emergency weapon intended for mass production. Developed by an engineer in the besieged city of Leningrad, it was constructed entirely of sheet metal plates that were welded,

known to German soldiers as the Erma (not the 'Schmeisser'). The MP 38 proved to be one of the most successful and influential submachine gun designs of World War II. It suited perfectly the offensive, mobile operations that were the credo of the Waffen-SS. What distinguished the MP 38 from earlier German submachine guns was that it was specifically designed for mass production, and thus used die-cast or stamped metal components wherever possible.

The MP 40, introduced in 1940, was even quicker and cheaper to manufacture and was produced in great numbers. The MP 40 weighed 4.7kg riveted, bolted or pinned together. It was a crude and unsophisticated weapon, but it did the job and thus remained in production throughout World War II.

A variety of American submachine guns saw limited service with SS personnel during the war, including the Thompson M1928 'Tommy Gun' and the M3 'Grease Gun'. Many of the American pistols and submachine guns that had fallen into German hands were rounded up during the autumn of 1944 and allocated to the 150th Panzer Brigade, commanded by Otto Skorzeny, the head of the élite SS Commando Raiding Forces. Skorzeny had rescued Benito Mussolini from imprisonment on Gran Sasso mountain back in 1943, and had seized the Hungarian capital in October 1944 to prevent Hungary's defection to the Allies. The 150th Brigade, which included SS paratroopers and commandos, was intended for deception and sabotage operations in the American rear areas during the German Ardennes counter-offensive of December 1944. The High Command issued American small arms to SS commandos who were disguised as American military police. They were tasked with changing road signs around, misdirecting traffic and generally sowing confusion and disorder in the American rear. Unfortunately, the employment of the brigade did not have the success envisaged. While teams from the brigade disguised in American MP uniforms did misdirect and confuse the panicked American troops on a number of occasions, most of the teams were eventually captured while still wearing American uniforms (they wore their own uniforms underneath) and executed for masquerading as Allied soldiers.

A plethora of both German and foreign small arms thus saw service with the Waffen-SS, belying the legend of the SS as an impressively equipped élite. Such variety of weaponry inevitably exacerbated problems of re-supply and the maintenance of combat effectiveness. Nevertheless, that so many Waffen-SS formations fought so well with such diverse ordnance underscores both the organisational flexibility of the Waffen-SS as well as the variety of its wartime service.

MP 38

German Designation: 9mm Maschinenpistole (MP) 38
Weapon Type: Submachine Gun
Calibre (Cartridge): 9mm (0.35in) (Parabellum)
Magazine Capacity: 32-round box

Length: 83.3cm (2ft 9in) extended
Weight: 4.1kg (9lb)
Muzzle Velocity: 390m/s (1279ft/s)
Rate of Fire: 500rpm (cyclic)

CHAPTER 2

Infantry Support Weapons

The widespread use of infantry support weapons provided Waffen-SS grenadiers with the direct and indirect fire support necessary to prevail both on the attack and in defence on the battlefields of World War II.

Infantry support weapons provide the rifleman with both direct and indirect fire support on the battlefield. The machine gun emerged in the late nineteenth century and tipped the balance between offence and defence towards the latter, which precipitated the inconclusive and costly trench warfare on the Western Front during World War I.

The light machine gun evolved in response to the unique tactical problems encountered amid the trench warfare on the Western Front between 1914 and 1918, where tripod-mounted heavy machine guns proved lethal but were of very limited mobility. What the Germans required, they

LEFT: *A 15cm Nebelwerfer 41 on the Eastern Front in 1943. This six-barrelled rocket launcher was mounted on the carriage of the 3.7cm Pak 35/36.*

realised, was a portable bipod-mounted machine gun that could accompany attacking infantry advancing into enemy trenches. In response, during 1915 the Germans designed the 7.92mm Maxim Maschinengewehr (MG) 08/15 light machine gun. This was a conversion of the standard Maxim MG 08 heavy machine gun fitted with a butt, simple bipod, smaller water jacket and pistol grip. Though heavy, it proved effective, and the inter-war German Army retained large numbers of them. They subsequently served with second-line Waffen-SS units throughout World War II. They equipped the SS Heimwehr *Danzig* during the September 1939 Polish Campaign, and as late as September 1944 the gun fought with the SS training units stationed in the Netherlands that saw action in the Arnhem fighting against Allied airborne forces.

ABOVE: *An excellent view of a 7.9mm MG 34 on a Dreifuss anti-aircraft tripod mounting. The ammunition feeder carries a Karabiner 89k rifle.*

The MG 08/18 followed in 1918. This was a much lighter, air-cooled weapon, and one which shaped post-war development. Very few entered service, though, before the war ended. Forbidden by the 1919 Treaty of Versailles from developing machine guns, the inter-war German military evaded this restriction by developing weapons outside Germany. This design work produced the air-cooled Dreyse MG 13 light machine gun, which became standard issue in the early 1930s. Limited in resources, the Reichswehr (the Weimar Republic German Army) realised that having distinct light and heavy machine guns was inefficient,

and thus German engineers evolved the revolutionary concept of a dual-purpose machine gun, one that could be used both as a heavy machine gun when tripod-mounted and in a light role when equipped with a bipod. This radical notion led to the evolution of the two finest machine guns of World War II – the MG 34 and its superlative successor, the MG 42.

THE REVOLUTIONARY MG 34

The MG 34 was the world's first general-purpose machine gun that could be utilised in both a light and heavy role. Of comparatively lightweight construction, the MG 34 fired either 'saddle' magazines or 50-round belts. The former consisted of two drums that fitted either side of the barrel. The magazine was spring-loaded, and rounds

32

were fed alternately into the gun chamber from each magazine. The belts consisted of 50 rounds lightly held together with steel clips, which allowed easy removal of a jammed round. Because of its dual role, the MG 34 had an air-cooled barrel ventilated by round holes in the sleeve, rather than the heavy and cumbersome water jacket used to water-cool previous heavy machine guns. In its heavy role the MG 34 was, therefore, lighter and more manoeuverable than its predecessors. The drawback of air cooling, however, was that it was not as efficient as water cooling and consequently the barrel of the MG 34 rapidly overheated with sustained fire. To counter this, the weapon was designed to allow rapid change of the barrel. The MG 34 remained in production from 1934 until 1945, though it was increasingly supplemented later in the war by the MG 42. In its light role the MG 34 weighed 11.5kg (25.4lb), which was much lighter than the 62kg (136.7lb) MG 08. The MG 34 possessed a wooden shoulder stock, a pistol grip and a V-notch rear sight. It had an effective range of 2000m (2188yds) and an impressive theoretical rate of fire of 800–900 rounds per minute. This weapon could deliver a formidable volume of fire, and Germany's enemies came to treat it with great respect. The gun was of solid and rugged design that stood the test of combat experience and the harsh weather conditions that German troops encountered, which ranged from Arctic cold to desert heat.

In 1940 Mauser began development of a cheaper and easier-to-produce successor. The new gun combined the best features of the MG 34 with experience gained from the MP 38/40 submachine gun in terms of ease and speed of manufacture. The resulting weapon – the MG 42 – was constructed largely from cheap and easy-to-produce die-cast and stamped components. Nonetheless, there was no sacrifice in quality; on the contrary, the MG 42 was not only the finest machine gun of World War II, it was one of the finest machine guns ever produced. The gun was simply superb: it had excellent handling qualities and was rugged and reliable. It had a tremendous rate of fire – up to 1550 rounds per minute – and produced a distinctive staccato 'brrpp' on firing that was unmistakable. Like the MG 34, the gun also possessed a quick-change barrel, yet barrel

Maschinengewehr 34

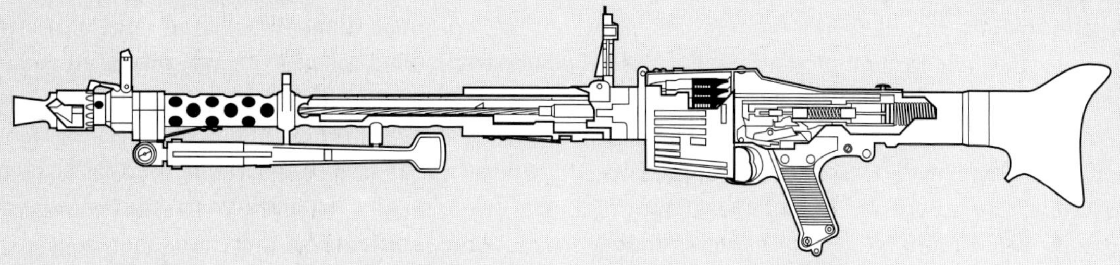

German Designation: 7.92mm Maschinengewehr (MG) 34

Weapon Type: Machine Gun

Calibre: 7.92mm (0.312in)

Magazine Capacity: 50-round belt or magazines

Length: 1219mm (48in)

Weight: 11.5kg (25.4lb)

Muzzle Velocity: 762m/s (2500ft/s)

Rate of Fire: 800-900rpm (cyclic)

Maschinengewehr 42

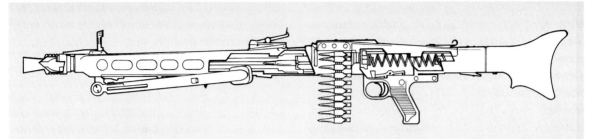

German Designation: 7.92mm Maschinengewehr (MG) 42
Weapon Type: Machine Gun
Calibre: 7.92mm (0.312in)
Magazine Capacity: 50- or 250-round belts

Length: 1.2m (3ft 11in)
Weight: 11.6kg (25.6lb)
Muzzle Velocity: 820m/s (2689ft/s)
Rate of Fire: 1550rpm (cyclic)

changing was even quicker than on the MG 34 and required only a single spring release. The MG 42 weighed the same as the MG 34 and it used the same belt or saddle magazines, allowing complete interchangeability of the two guns in combat. As with its predecessor, individual belts could be clipped together for more sustained fire. Both guns were also fitted with a muzzle-flash suppressor which, when combined with flashless powder, made it difficult to detect the machine gun from its muzzle-flash.

Allied soldiers on all fronts quickly came to appreciate and fear the infamous 'Spandau' as it was dubbed. American wartime intelligence concurred and assessed the gun as 'the most remarkable gun of its type ever developed', an impressive testimonial from an enemy! Further testament of the quality and durability of the weapon is that it continues to see service, albeit in slightly modified form, as the MG 3 of the German Army today. It is, in fact, one of the few weapons of World War II that has not been rendered obsolete by technological advances. The quality of German machine guns lay at the core of the formidable defensive reputation that the Waffen-SS gained during the

latter stages of World War II. The Germans developed effective defensive tactics and techniques designed around the prowess of the MG 34 and MG 42. Indeed, Waffen-SS troops understood that they had considerable defensive staying power against enemy infantry as long as they could keep their machine guns operational and deployed with good fields of fire. Waffen-SS grenadiers thus took great pains to deploy their machine guns in the most advantageous defensive positions and employed elaborate camouflage to cover them. When time and conditions allowed, machine-gun crews prepared several fall-back positions. Appreciating the value and effectiveness of their machine guns, Waffen-SS troops were loath to risk them unnecessarily and if unsuccessful at halting an enemy advance, machine gunners had standard operating procedures – Hitler's 'Stand Fast' orders notwithstanding – to retire to prepared rearward positions and resume the battle before the enemy got too close. As long as SS grenadiers could keep the majority of their machine guns intact and operational, they could hold up attacking infantry many times their number. Indeed, a couple of well-sighted, well-hidden and well-supplied MG 42 machine guns could hold up an entire attacking regiment for hours on a frontage of several miles. Such was the importance of machine guns to German defensive success that their crews were chosen with great care, and in reality the nine-man SS grenadier squad essentially existed to service its machine gun. The primary gunner – Schütze 1 – was invariably the most experienced and decorated grenadier in the squad, although it was also a great benefit if he was burly too because he had to carry the gun. His team mate, Schütze 2, fed the ammunition belts and saw that the gun remained operational by undertaking periodic barrel changes and the clearing of jammed rounds. Two men – generally the least experienced recruits – did nothing but bring up fresh ammunition for the gun. The rest of the squad generally deployed in foxholes to cover all possible approaches to the machine-gun nest. The great advantage of this system was that German infantry squads could sustain very heavy losses – up to two-thirds – but remain effective if they could keep their machine gun functional. As long as the squad had a gunner, loader, and a munitions carrier it could sustain almost the same volume of fire as a full squad. This reality explains why so often SS infantry squads were able to continue resistance after suffering losses that would have rendered Allied infantry ineffective.

STORM OF STEEL

Illustrative of the defensive power the MG 42 accorded was the action southwest of Caen, astride the Odon River, conducted by the 6th Company, 1st SS Panzergrenadier Regiment, 1st SS Panzer Division *Leibstandarte Adolf Hitler* on 28 June 1944. SS-Schütze (Private) Erich Göstl found his squad facing an entire British battalion assault. Despite being shot and blinded in one eye Göstl refused to relinquish his gun and such was his withering stream of fire which scythed through the cornfields that he pinned down the entire attacking battalion. Even when he was hit again, this time by shrapnel that damaged his other eye, Göstl stayed at his post and fired belt after belt of munitions at the enemy. His actions caused the entire British attack to stall and the 'Tommies' retired to their start line. For his outstanding heroism, Göstl was cited in dispatches and awarded the Knight's Cross. His citation fittingly described the man as 'heroic without exaggerated pathos, a man of deeds rather than words'.

Light machine guns were also very powerful offensive weapons, as SS-Scharführer (Senior

BELOW: *An SS mortar crew about to fire an 8cm GrW 34 medium mortar. The loader is about to drop the 3.5kg (7.7lb) round down the tube.*

Sergeant) Johann Fiedler demonstrated on 21 April 1942, while serving with the *Totenkopf* Division at Demyansk on the central sector of the Eastern Front. When superior Soviet forces pinned down his company and it faced destruction, Fiedler single-handedly charged the enemy positions with his MG 34 in its light role and, despite being wounded, stormed the Soviet trenches, took 39 prisoners and captured five

machine guns and an anti-tank gun. For his determination and bravery he was awarded the Knight's Cross, one of the first SS NCOs to receive this coveted award.

Despite, or perhaps because of, the quality of the MG 34 and MG 42, German production never came close to matching demand, and throughout the war the SS had to supplement these weapons with a wide range of captured machine guns.

leGW 36

German Designation: 5cm leichte Granatwerfer (leGW) 36
Weapon Type: Light Mortar
Calibre: 5cm (2in)
Length (of barrel): 46.5cm (1ft 6in) [=L/9.3]
Weight: 14kg (30.9lb)
Bomb Weight: 0.9kg (2lb)
Muzzle Velocity: 75m/s (296ft/s)
Maximum Range: 520m (596yds)
Rate of Fire: 15-25rpm

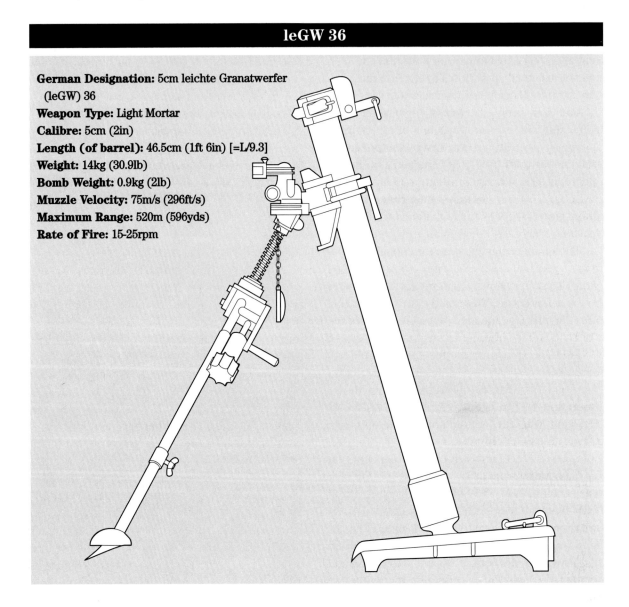

This was particularly so in the late 1930s when the army opposed SS expansion. During this period, the fledgling Waffen-SS acquired small quantities of rechambered Knorr-Bremse Swedish 6.5mm LH33 machine guns, which were designated in German service as the MG 35. Only Waffen-SS forces used this machine gun during the early war years. During 1939–40 two Czech machine guns, the ZB vz 30 and 37, designated in German service as the MG 26(t) and 30(t), also saw widespread service among the Waffen-SS. These two machine guns were the inspiration for the British Bren gun and Besa light machine gun, respectively. What made these two guns valuable was that they fired the standard German 7.92mm cartridge. The ZB vz 30 fired top-mounted 20- or 30-round box magazines and was cocked by means of a prominent handle on top of the barrel, forward of the breech. The gun weighed 10kg (22lb) and had a cyclic rate of fire of 500 rounds per minute and an effective range of 2000m (2188yds). It served with the *Totenkopf* and *SS-Polizei* Divisions in the 1940 campaign in the West.

The ZB vz 37 was a heavy, air-cooled machine gun on a substantial tripod mounting. It weighed 19kg (41.8lb), had a maximum rate of fire of up to 700 rounds per minute and an effective range of 2000m (2188yds). Large numbers of both these Czech machine guns saw service with the SS Brigade *Nord* raised from former SS Totenkopf concentration guard personnel and deployed on the Finnish sector of the Eastern Front during Operation 'Barbarossa'. They also served in the SS *Prinz Eugen* Volunteer Mountain and *Florian Geyer* Cavalry Divisions.

Soviet machine guns in German use

Three main types of Soviet machine guns served with the Waffen-SS: the 7.62mm Degtyzrev DP light machine gun, the light 7.62mm Goryunov SG 43, and the heavy 12.7mm Degtyzrev DshK 38. They were widely used because the Germans captured millions of rounds of ammunition for these weapons during the early stages of Operation 'Barbarossa'. The Degtyzrev DP, which entered service in 1938, was a simple weapon of excellent quality – it contained just six moving parts! Weighing 11.9kg (26.2lb), it featured a wooden stock, rifle (later pistol) grip, ventilated barrel sleeve, flash suppressor, integral bipod mount and a top-loaded drum magazine that held 47 rounds. The gun was gas operated and had a rate of fire of up to 600 rounds per minute, though effective range was only 800m (875yds). The Goryunov SG 43 was also belt fed and was normally mounted on a wheeled carriage, with or without a splinter shield. The basic gun weighed 13.8kg (30.4lb) but the carriage and shield added a substantial 23.1kg (50.9lb) to the weight. The gun was fed by long 350-round belts and had a rate of fire of 250 rounds per minute and an effective range of 2000m (2188yds). The heavy DshK 38 belt-fed machine gun was of similar performance to the American Browning 12mm heavy machine gun on which it was based.

Mortar firepower

The modern infantry mortar first appeared on the battlefields of World War I. It combined firepower with excellent mobility to provide indirect fire support at short range. A mortar is essentially a high-elevation, smooth-bore weapon that fires a fin-stabilised bomb on a high plunging trajectory. The firing tube is mounted on a baseplate and there is normally a bipod for support which is adjusted to change the elevation/trajectory and hence the range. Mortar bombs contain a propellant charge at the bottom which is ignited by the simple process of dropping the bomb into the barrel of the mortar so as to strike the firing pin positioned at the bottom. Because it was rugged, cheap and simple to manufacture and use, the mortar proved particularly valuable in the trench warfare on the Western Front. Invented by the British, the Germans remained slow to realise its potential. It was only the extensive post-war

Verfügungs, *Totenkopf* and *SS-Polizei* Divisions in the 1940 campaign in the West, but the Germans terminated production of the leGW 36 in 1941 because the mortar was too complex, too expensive and lacked punch.

THE leGW 36 IN ACTION

From 1942 the leGW 36 was progressively phased out of frontline service but continued to serve with SS foreign volunteer, rear-area and training units until the end of the war. The Croatian SS Volunteer Mountain Division *Handschar*, raised during 1943 in the Balkans, for example, possessed light mortar platoons in each of its mountain infantry companies, which were equipped with three of the mortars. Similarly, the *Landstorm Nederland* security force raised in regimental strength in Holland fielded a mortar platoon equipped with three 5cm light mortars in each of its nine infantry companies as late as the summer of 1944. Minor elements of this formation became embroiled in the bitter fighting at Arnhem in September 1944 during Montgomery's 'Market Garden' Offensive, the 5cm leGW 36 once again being used against Allied troops. However, the remaining mortars were progressively withdrawn during the autumn and winter of 1944–45 as the German High Command upgraded the force to divisional status in February 1945, to become the 34th SS Grenadier Division *Landstorm Nederland*.

The largest and rarest mortar to see operational service with the Waffen-SS was the 12cm schwere Granatwerfer 42 (sGW 42). This was a deadly heavy mortar, which – fortunately for the Allies – remained a rare weapon. Developed during 1942–43 in response to encounters with the heavy Soviet mortar of the same calibre, the Germans designed a virtual copy of the Red

analysis that illuminated the importance of the weapon and the Germans vigorously explored mortar development between the wars. By 1939 the mortar had become a standard infantry support weapon in all Western armies giving infantry – both in attack and defence – a valuable high-explosive capability beyond the range of rifles or hand grenades. The drawback of the mortar, however, was a general lack of accuracy, it being an area weapon. Even with an experienced mortar team, it generally required 10 bombs to achieve a direct hit on a target.

During World War II, the Germans enjoyed considerable standardisation in mortar types with three basic weapons, though once again production shortfalls ensured that a range of foreign mortars served with the Waffen-SS. The first German mortar to enter service was the 8cm Granatwerfer 34 (8cm GW 34), which actually was a 8.1cm-calibre weapon despite its designation. This weapon remained the standard German infantry mortar throughout World War II. It fired a 3.4kg (7.48lb) bomb to a maximum distance of 2400m (2626yds), almost five times the range of its smaller cousin, the 5cm leichte Granatwerfer 36 (leGW 36) which entered service in 1936. The leGW 36 fired a small 0.9kg (1.96lb) charge to a maximum range of only 500m (547yds). Both mortars fought with the *SS-*

Army weapon. The 12cm sGW 42 fired a 15.8kg (34.8lb) bomb to a range of 6050m (6619yds), an impressive range for a mortar. Introduced in 1943, it was delivered to a select few SS units in 1944 but was never common. None of the SS divisions stationed in the West, for example, had received the mortar prior to the onset of the June 1944 Normandy Campaign. The Germans intended to issue the weapon to the heavy companies of SS panzergrenadier and grenadier regiments in lieu of medium mortars, but this rarely materialised, though the 9th SS Panzer Division *Hohenstaufen* received 24 during the summer of 1944. Instead the mortar was often retained at corps or army level. Indeed, Himmler intended to raise motorised heavy mortar companies for each SS corps, but few materialised. During the spring of 1944, for example, II SS Panzer Corps

then forming in the West was authorised two heavy motorised companies, each equipped with a dozen 12cm mortars, but these were never formed due to production shortages. Rather, when the 12cm sGW 42 did serve in SS formations, it generally did so in the regimental cannon companies of grenadier and panzergrenadier regiments, in place of infantry guns. In fact, the 12cm sGW 42 was cheaper, quicker-to-produce, more mobile and had superior performance to infantry guns. The mortar was therefore much liked by its gunners.

Throughout the war the Waffen-SS made extensive use of mortars, more so than its opponents.

BELOW: *Probably a battery of heavy 30cm Raketenwerfer 56 rocket launchers in action on the Eastern Front. This weapon was rare in SS use.*

That reliance increased as the war progressed because mortars were cheap and easy to produce and because they were mobile and easy to camouflage, which made them less vulnerable to air attack than artillery or infantry guns as Allied aerial superiority increased later in the war. Indeed, the quality and availability of mortars powerfully reinforced the defensive firepower that German machine guns provided. Waffen-SS grenadiers habitually pre-registered the fire of their mortars, not only on the forward battle zone but also on their own defensive positions and machine-gun nests – another reason why SS

BELOW: *Loading a 10-barrelled 15cm Maultier Panzerwerfer 42 rocket launcher, autumn 1944. The launcher is mounted on the rear of a halftrack.*

machine gunners were prepared temporarily to relinquish their forward positions to the enemy while they reorganised and summoned reinforcements to counterattack.

This fate befell the riflemen of the 3rd Canadian Infantry Division during its attack on Carpiquet airfield west of Caen on 8 July 1944. The defenders of the 26th SS Panzergrenadier Regiment, 12th SS Panzer Division *Hitlerjugend*, had established a strong defensive position and had carefully registered the fire of their 8cm GW 34 mortars on the forward battle zone and their own outpost line. That morning the Canadian infantry confidently advanced behind a massive, rolling artillery barrage that temporarily suppressed German defensive machine-gun fire. But in the main line of resistance, SS mortar crews

Nebelwerfer 42

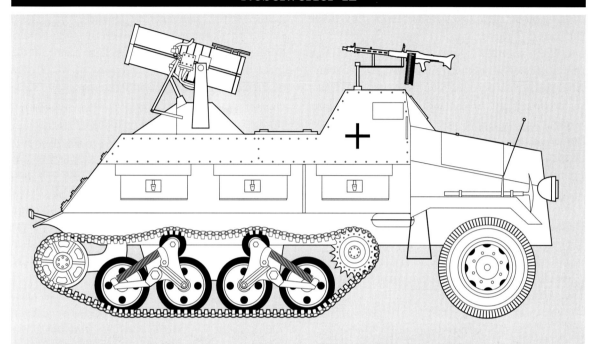

German Designation: 21cm Nebelwerfer (NbW) 42
Weapon Type: Multi-barrelled Rocket-Launcher
Calibre: 21.4cm (8.4in)
Length (of barrels): 1.3m (4ft 3in)
Weight: 1100kg (2448lb) = 1.1 tonnes (1.1 tons)

Rocket Weight: 112.6kg (248lb)
Initial Velocity: 320m/s (1049ft/s)
Maximum Range: 7850m (8588yds)
Rate of Fire: 5 rockets in 8 seconds; 3 salvos x 5
  rockets in 5 minutes

held their fire until the Canadian artillery began to lift and then poured down dozens of rounds, making the Canadian riflemen think that their own artillery had fallen short, which sent them into panic. One Canadian battalion was so disrupted it could not participate in the attack at all and the entire Canadian timetable was disrupted. Consequently, it took all day for the vastly superior Canadian forces to force the young Hitler Youth soldiers out of Carpiquet village, but the airfield remained firmly under SS control.

The Waffen-SS pressed into service a vast array of captured mortars. The most commonly used foreign model in SS service were variations

of the basic 8.14cm Stokes-Brandt Mortar, the most widely used mortar in Europe at the start of the war. The French Army deployed over 8000 alone in 1940, and the Germans captured additional examples from Austria, Poland, Holland, Denmark and Yugoslavia as well. These were issued in lieu of German medium mortars and generally served with SS garrison and training units. The SS *Landstorm Nederland*, for example, fielded some 18 8.14cm Granatwerfer 275(d) mortars, which were Danish license-built versions of the Stokes. This regiment also had three 1935-model French 6cm mortars, designated in German service as the 6cm Granatwerfer 225(f), another light French mortar that made its way into SS service in limited numbers.

As noted above, SS formations deployed on the Eastern Front also used captured Soviet 8.2cm and 12cm mortars. These had ranges of 3100m and 5000m (3391yds and 5634yds), respectively. After the retreat across France in August 1944, the decimated 102nd SS Rocket Launcher Battalion re-equipped one of its batteries with a dozen ex-Soviet 12cm mortars. These provided sterling fire support against the British airborne forces penned in the Arnhem-Oosterbeek enclave during 'Market Garden'. The Serbian Volunteer Corps, formed from Russian Tsarist émigrés in the Balkans and which the Germans administratively subordinated to the SS during the spring of 1945, also possessed a mortar platoon of four ex-Soviet 12cm mortars in each of its battalion-level heavy weapons companies.

ROCKET TECHNOLOGY

Rockets are the oldest known form of artillery, the Chinese having used them centuries before the western development of gunpowder cannon in the Middle Ages. Today, rockets are an integral feature of most modern militaries. However, during the nineteenth and early twentieth centuries, they were not widely employed by European nations. Consequently, the 1919 Treaty of Versailles, which prohibited Germany from developing heavy artillery, said nothing whatsoever about restricting German rocket development. Denied the right to develop heavy artillery, the treaty spurred the Germans into developing rocket weapons. Camouflaged under the designation of Nebelwerfer (literally 'smoke projector'), between the wars the Germans developed the first effective modern rocket systems as General Walter Dornberger's scientists explored the possibilities of rocketry at a secret testing centre located at Peenemünde on the Baltic coast.

One of the most fundamental problems with rockets that had curtailed their previous use was

that placing the propellant charge at the rear of the rocket produced in-flight instability which made them inaccurate. During the 1930s Dornberger's team came up with the then revolutionary idea of putting the propellant charge at the front of the rocket, the back-blast of which was exhausted through venturis at the rear of the rocket astride the high-explosive charge. By angling these venturis it was possible to impart spin to the rocket, creating spin-stabilisation that enhanced the weapon's range and accuracy. The result was an effective rocket delivery system that was perfected as Germany went to war in 1939. In 1940 a rocket-launcher emerged, the 15cm Nebelwerfer 41 (NbW 41), which fired a range of multiple 15cm spin-stabilised rockets that carried a high-explosive, smoke and poison gas warheads. Hitler wisely refrained from employing the latter rounds, however, remembering his own bitter experience of being gassed and partially blinded during World War I. Fully loaded the NbW 41 weighed 770kg (1694lb) and fired six 34kg (74.8lb) Wurfgrenate 41 rockets to a maximum range of 6900m (7553yds). It took 10 seconds to fire a full salvo, but since the launcher had to be manually reloaded it could fire only three salvoes in five minutes. The launcher was fitted on a slightly modified 3.7cm Pak 35/6 anti-tank gun carriage.

SS ROCKET-LAUNCHER BATTALIONS

This rocket launcher began to be issued in 1941 to a new, élite rocket-projecter branch (Nebeltruppe) and served in independent army rocket-launcher battalions, and later regiments and brigades, on all fronts until the end of the war. The Nebeltruppe tenaciously resisted the grasping hands of the Waffen-SS, and it was not until 1943 that Himmler was able to secure a small proportion of new production. This allowed the

RIGHT: *A 15cm Panzerwerfer 42 Maultier rocket launcher. The white 'F' on the left of the vehicle indicates it belongs to the* **Frundsberg Division.**

sIG 33

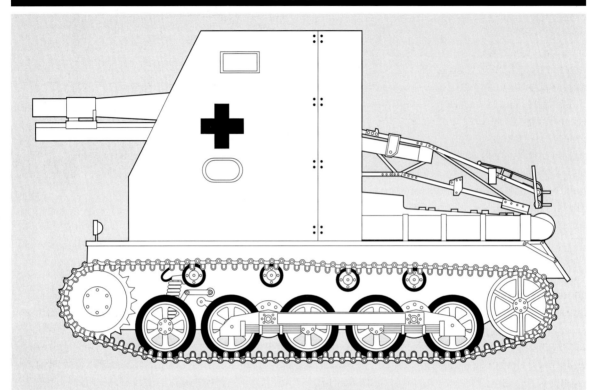

German Designation: 15cm Schweres
 Infanteriegeschütz (sIG) 33
Weapon Type: Heavy Infantry Gun
Calibre: 14.9cm (5.9in)
Length (of piece): 1.7m (5ft 7in) [= L/11.4]
Weight: 1800kg (3969lb) = 1.8 tonnes (1.8 tons)

Traverse: 11 degrees
Elevation: -4 to +75 degrees
Muzzle Velocity: HE = 240m/s (787ft/s)
Maximum Range: 4700m (5142yds)
Rate of Fire: 2-3rpm

SS to raise its first rocket-launcher battalions during that year equipped with the fearsome 15cm NbW 41.

During late 1942 a much heavier rocket launcher, the 21cm (8.3in) Nebelwerfer 42 (NbW 42), entered service, though it would never be as common as its smaller cousin. When fully loaded it weighed 1100kg (2420lb) and had five barrels, instead of six, to keep weight down. Each rocket weighed 113kg (248.6lb) and had a maximum range of 7850m (8588yds). The NbW 42 could fire a full salvo in eight seconds and three full salvoes in five minutes. Essentially a scaled-up version of the NbW 41, it was mounted on the same carriage. Hastily designed to rush it into production, it had a more conventional design, however, and its Wurfgrenate 42 (WGr 42) round proved less accurate than the Wurfgrenate 41.

Another rocket system developed was the dual-calibre 28/32cm Nebelwerfer 41 which could fire both 28cm and 32cm Wurfkorper Flamm incendiary rockets. Internal guide rails also allowed the 21cm WGr 42 rocket to be fired from this launcher, making it a versatile weapon. This system had drawbacks, however. The rockets had the warheads mounted at the front and were therefore inaccurate and suffered from short range: only 1925m (2106yds) and 2200m (2407yds) for the 28cm and 32cm rocket respectively.

Though designed primarily as an anti-personnel weapon, these rockets proved lethal against open or soft-skinned vehicles and the blast effect was so great that a direct hit could severely concuss a tank crew. The first two SS corps rocket-launcher battalions, numbered the 101st and 102nd, came into existence during 1943, and by the spring of 1944 they had been joined by the 103rd, 104th and 105th. In September 1944, these units were renumbered the 501st–505th as part of general reclassification of SS corps troops. The standard battalion organisation consisted of four batteries of six 15cm rocket launchers, a total of 24 launchers and 144 barrels, though organisational establishments varied. The 101st SS Rocket-Launcher Battalion, for example, had one battery equipped with six 21cm rocket launchers. During the spring of 1945, Himmler intended to concentrate the existing SS rocket-launcher battalions into a brigade. Though the headquarters of the 1st SS Rocket-launcher Brigade came into existence, the German collapse in the last months of the war ensured that this project never came to fruition.

A FORCE SPREAD TOO THIN

Rocket launchers also appeared less frequently at divisional level. Indeed, Himmler originally intended that each of the seven premier Waffen-SS panzer divisions should have a rocket-projector battalion, but this could not be realised. Only the 1st and 12th SS Panzer Divisions received such battalions. The 12th SS Rocket-Launcher Battalion, formed in France during 1943, deployed four batteries each of six 15cm NbW 41. These weapons were all lost in the bitter defensive fighting experienced in Normandy, the last launchers

ABOVE: *The crew of a 15cm sIG 33 heavy infantry gun of the* Wiking *Division ready their weapon for action on the Eastern Front during the summer of 1942.*

being destroyed in the Falaise Pocket during mid-August 1944. The remnants withdrew to Cologne for reconstitution and fought with the *Hitlerjugend* Division in the Ardennes Offensive and then in Hungary. Out of launchers and rockets, the battalion fought as infantry in Austria during the last weeks of the war.

Only a few other SS formations fielded rocket-projector units. Among these was the ad hoc SS Panzer Brigade *Gross* which was raised in mid-August 1944 at the Seelager SS proving grounds area near Riga in Latvia. The brigade contained a heavy motorised rocket-launcher platoon equipped with two 21cm NbW 42. It helped to parry the Red Army advance towards Memel and contributed to the successful German retreat into the Courland enclave in Latvia during the autumn of 1944.

The rarest rocket launcher to see service with the Waffen-SS was the super-heavy 30cm Raketenwerfer (RW) 56, which may have served with just a single SS unit. This system, which entered service in 1944, consisted of a 30cm Wurfkorper 127kg (279.4lb) rocket fired from the racks of the 28/32cm launcher. Mounted as a six-pack, the RW 56 possessed a range of 4550m

(4878yds) but lacked accuracy. It could fire a full salvo in 10 seconds and two salvoes in five minutes. It saw service with the 32nd SS Heavy Rocket-Launcher Battery which was raised on 4 February 1945 at Lübbinchen from the SS Rocket-Launcher Training and Replacement Battalion which was then field testing the first three 30cm rocket launchers delivered to the Waffen-SS. These three launchers plus a single 21cm rocket launcher joined the 32nd SS Grenadier Division *30 Januar* then forming at the Kurmark proving grounds. The battery fought on the Oder Front with the division during February 1945 in a vain effort to prevent the Soviets consolidating their bridgeheads on the west bank of the River Oder. Once the battery had exhausted all its available munitions, however, it was dissolved at the end of February.

MULTI-BARRELLED FIREPOWER

The vielfachwerfer multi-barrelled rocket launcher again epitomised the wartime rivalry that existed between the Waffen-SS and the German Army. In the face of bitter opposition from the Army's artillery branch to Waffen-SS attempts to acquire rocket launchers, the SS ultimately developed its own fin-stabilised rocket system in competition to the Army's Nebelwerfer units. During 1941, the opening year of the war in the East, SS troops came into contact with the Soviet RS-82 'Katyusha' rocket launcher. The SS captured several Katyushas intact and shipped them back to Germany for examination. Himmler was so impressed with the weapon that he ordered SS factories to build a slightly modified German version of the weapon, the 8cm Raketen-Vielfachwerfer. The launcher comprised two dozen launching rails, each capable of firing two missiles, for a total salvo of 48 rockets. The weapon was capable of 360 degree traverse and could elevate through 37 degrees. Each rocket weighed 6.9kg (15.2lb) and had a maximum range of 6300m (6897yds). The launcher was typically mounted on ex-French Army Somua armoured halftracks and served exclusively in Waffen-SS rocket-launcher units. Since the SS produced the Vielfachwerfer as a private initiative from its own resources without the involvement of the Army Weapons Department, production was inevitably very limited.

The first SS unit equipped with these rocket-launchers was the 101st, later 522nd, SS Vielfachwerfer Battery, which was attached to the I SS Panzer Corps during early 1943. The battery fielded six of the 8cm rocket launchers. During the spring of 1943, a second battery, the 521st, was activated at Lanlingen and equipped with four 8cm launchers. It deployed to Kharkov on the Eastern Front, where it came under control of the 502nd SS Rocket-Launcher Battalion. It fought with I SS Panzer Corps during the recapture of Kharkov and then in the bitter attritional struggle at Kursk. During the retreat to Poltava during the summer of 1943, the battery fought alongside the 5th SS Police Regiment where it lost its remaining launchers. It then returned to Langlingen for replenishment in October 1943.

FIGHTING TO THE BITTER END

The 521st battery was back in action in the spring of 1944 on the northern sector of the Eastern Front, attached to III Germanic SS Panzer Corps. It fought on the Narva River but suffered heavy losses during the retreat from the Baltic states to Courland during the late summer of 1944 and was disbanded in early November. Himmler reconstituted the battery at the SS proving grounds at Kurmark during January 1945, and the next month the unit fought on the Oder Front attached to the 506th SS Rocket-Launcher Battalion, the last SS rocket launcher unit to be formed. It participated in the bitter fighting retreat in front of Berlin during April 1945, and disintegrated amid the final German collapse on the Eastern Front during the last two weeks of the war.

The Waffen-SS also made use of the infantry gun, another German innovation of the inter-war period. An infantry gun was a lightweight, mobile support weapon operated by infantry and capable of giving immediate and intimate indirect fire support beyond the range of mortars whenever it was needed by the grenadiers. During World War I, the Germans had generally assigned light field guns or pack howitzers to dedicated infantry fire support, though they remained under artillery control. During the 1920s considerable debate materialised concerning the relative merit of 7.5cm field guns or heavier 10.5cm light howitzers. In the German military, the larger piece ultimately won out and the light field howitzer became the standard divisional artillery weapon by the late 1930s. This left the infantry ill-supported, and their vigorous objections to the switch ultimately led to the development of an entirely new category of weapon that the Germans termed infantry guns.

In response to various infantry demands, the Germans developed two basic designs in the early 1930s: a 7.5cm light and a 15cm heavy infantry gun to equip new cannon companies of infantry regiments. In September 1939, each of the four SS motorised infantry regiments fielded a single regimental gun company equipped with eight towed 7.5cm light infantry guns. Delays in production of the heavier 15cm gun ensured that it did not enter Waffen-SS service until 1940. The 7.5cm leichte Infanteriegeschütz 18 (leIG 18)

BELOW: *The crew of a 7.5cm leIG 18 light infantry gun belonging to the* **Leibstandarte** *Division march behind their horse-drawn gun on the Eastern Front.*

entered service in 1929. It weighed 400kg (882lb) and fired a 6kg (13.2lb) shell to a maximum range of 3375m (3692yds). It remained the standard German light infantry gun throughout the war.

In 1944 small numbers of a new gun, the 7.5cm leIG 37 began to supplement the leIG 18. This weapon combined the barrel of the prototype leIG 42 infantry gun with existing 3.7cm Pak

35/36 anti-tank gun carriages, then available in numbers. The leIG 37 weighed 510kg (1125lb) and had a range of 5150m (5634yds). Only very limited numbers are believed to have reached the Waffen-SS late in the war, however. A dozen fought with the 9th SS Panzer Division *Hohenstaufen* during the September 1944 fighting at Arnhem. The gun's design reflected the increasingly desperate and hasty improvisation that characterised German weapons procurement during the latter stages of the war.

Production of the 15cm (5.9in) schwere Infanteriegeschütz 33 (sIG33) heavy infantry gun commenced in 1935, and after extensive trials the gun entered service in 1938. With a weight of 1700kg (3750lb), the sIG 33 was heavy for a infantry support weapon, and possessed the bulk of an artillery piece. From 1940 until the end of the war two 15cm SiG 33 guns formed the heavy infantry gun platoon of SS grenadier and panzergrenadier regiments. The gun fired a 38kg (84lb) shell to a range of 4700m (5140yds). Though it proved a reliable weapon, its weight and bulk proved significant tactical limitations. It remained in service until 1945, though it was never particularly common.

INFANTRY GUN TACTICS

Infantry guns could provide SS grenadiers with vital offensive and defensive fire support, particularly when artillery was unavailable, as was often the case on the sparsely defended Eastern Front. Waffen-SS gunners usually deployed their infantry guns in the rear of their 'Hauptkampflinie' – main defensive position – and coordinated their fire with more forward-deployed mortars.

Offensively, infantry guns could accompany attacking grenadiers and even provide direct fire if necessary. Since they were relatively easy and

quick to emplace and make ready, they were very useful in mobile operations, providing covering fire while artillery could be readied for action. This was precisely the situation faced by the 26th SS Panzergrenadier Regiment of the 12th SS Panzer Division *Hitlerjugend* on 7 June 1944 on its 'Marsch zum Front' to repel the Allied D-Day landings. Each of its three battalions were reinforced by a platoon of two heavy infantry guns. The division's scouting forces, which had just reached the Ardennes Abbey, west of the city of Caen, witnessed the Canadian infantry of the Regina Rifles advance together with the Sherman tanks of the 27th Canadian Armoured Regiment. Without waiting for reinforcements or artillery support, the SS panzergrenadiers went straight into the counterattack from their assembly points, relying on surprise, élan and the covering fire of their 15cm infantry guns for support. These pieces provided valuable firepower for the SS counter-strike that routed the Canadian spearhead and sent it reeling backwards. They helped suppress the local defensive mortar fire the Canadians called down to cover their retreat, and ultimately it took the fire of every Anglo-Canadian artillery piece ashore, plus the guns of the Allied naval invasion armada, to bring the SS armoured counter-thrust to a halt.

FOREIGN INFANTRY GUNS IN SS USE

From the middle of the war, however, Germany switched productive resources away from infantry guns to higher priorities. Increasingly heavy mortars and field guns began first to supplement, and then to replace, infantry guns in regimental cannon companies. The Germans also pressed a variety of captured enemy light artillery pieces, particularly field guns, into service as infantry guns. The two foreign weapons that saw most extensive German service as light infantry guns was the Soviet 7.62cm (3in) M1936 and M1939 field guns that were captured in sizeable numbers during the first year of Operation

'Barbarossa'. The Germans quickly appreciated the robustness and versatility of these weapons, and issued them to divisional artillery regiments and regimental gun companies.

Another substitute infantry gun was the ex-Soviet 7.62cm Infantry Gun-Howitzer M1927, which was designated in German service as the 7.62cm Infanteriekanonenhaubitze (IKH) 290(r). It saw widespread service as an infantry gun, particularly among SS garrison forces in Western Europe. The 9th SS Panzer Division *Hohenstaufen*, for example, received 18 of the guns during the spring of 1943. Several guns originally intended for mountain warfare ultimately doubled as infantry guns. These included the 7.5cm Gebirgsgeschütz 283(j) and 285(j). These Czech-built ex-Yugoslavian mountain guns came into the possession of the SS in very limited numbers late in the war.

GENESIS OF THE FLAMETHROWER

The Germans introduced the modern flame-thrower in the Argonne fighting of October 1914. The basic elements of the flamethrower comprised tanks of separated inflammable liquid and compressed gas. When a trigger valve was pressed, the gas forced the liquid out along a firing tube where it ignited and ejected flame up to 35m (38yds). The Germans thus introduced a terrible and potent new weapon onto the modern battlefield. How and why the Germans developed this new weapon is still unclear, though they told a perhaps apocryphal story that when in 1914 the Kaiser made an unannounced visit to a training exercise on a hot summer's day, he found troops playfully drenching each other with water hoses. When the Kaiser enquired of the commander what his troops were doing, the quick-thinking officer replied that in wartime he intended to replace the water with petroleum and set the enemy on fire. The Kaiser was so impressed by the officer's apparent ingenuity and initiative that he ordered new flame projec-

tors to be developed and troops trained to use them. And hence, so the story went, flame was introduced as a weapon into the Imperial German Army!

TERROR WEAPON

The flamethrower remained a clumsy and complex weapon of particular effect against enemy infantry morale, but was never a common or significant system in its own right, having limited tactical effectiveness. Indeed, the early German flamethrowers were so bulky that they required up to three men to use, and therefore could not be widely employed in the mud of the trenches of Flanders. Nonetheless, the weapon left an indelible imprint on those fortunate enough to survive a flame attack. Indeed, flamethrowers are one of the most terror-inspiring direct-fire weapons of modern war, and a flame attack can so demoralise enemy infantry as to induce them to surrender rather than suffer incineration. Flame became a favoured weapon in the hands of German Stormtroop units in the very latter stages of World War I, and the Allied victors were sufficiently impressed to ban Germany from developing flame weapons in the 1920s. Despite these restrictions, however, German engineers tinkered with the basic concepts and ideas involved in flame-thrower construction.

Thus in 1935, when Hitler repudiated the Treaty of Versailles, the German military quickly developed the Flammenwerfer 35 (FlW 35) flamethrower. The impact of the inter-war development prohibitions were apparent, however, since the FlW 35 remained large, bulky and cumbersome. Intended for one-man use, it was too heavy for this and a crew of two generally man-handled it. With time and experience, the Germans progressively reduced the weight of their flamethrowers. The FlW 35, which remained in production until 1941, weighed 35.8kg (79lb) and became the first flamethrower to be used by SS engineers, though it remained rare in SS service. The fuel was carried in a large tank mounted on the soldier's back alongside a cylinder of compressed nitrogen gas. A single trigger valve on the projector both released the fuel and ignited it. The FlW 35 had a range of 25–30m (27.4–32.8yds), and its 11.8-litre (2.6-gallon) fuel tank provided approximately 10 seconds of fire. As protection against accidents or blow-backs, crews usually wore fire-protective jackets and leggings, as well as a simple plastic transparent face mask attached to the regular field helmet.

A smaller, more portable, version developed in 1940 weighed about one-third less. This was the Flammenwerfer klein verbessert 40 (literally the 'smaller better' flamethrower). On this weapon the fuel and nitrogen were arranged in two 'lifebuoy' rings with the outer ring containing the fuel, the inner the nitrogen. This method reduced weight and more importantly offered a better fit on the engineer's back, which made it easier to operate. This weapon was superseded the following year by an even lighter FlW 41 which weighed only 18.4kg (40.4lb). This flamethrower had two horizontal tanks of which the lower and larger contained the fuel. The weapon used the same trigger system as previously but a

Flammenwerfer 41

German Designation: Flammenwerfer (FlW) 41
Weapon Type: Flame Thrower
Weight: 22kg (48.4lb)

Fuel Capacity: 7 litres (1.55 gallons)
Range: 30m (33yds)
Burst Duration: 10 seconds

new hydrogen ignition system was introduced with a long thin cylinder to contain the hydrogen mounted over the projector.

Combat experience on the Eastern Front during the winter of 1941–42 demonstrated that the weapon's hydrogen ignition system was susceptible to the extreme cold of the Soviet winter. This led to the development of a new cartridge ignition system during 1942, which was combined with a new projector that had a magazine containing a feed system for 10 incendiary cartridges. Every time the trigger was depressed, a cartridge loaded, fired and ejected, igniting the fuel jet in the process. This new system, known as the Flammenwerfer mit Strahlpatrone 41, became the standard backpack German flamethrower until the end of the war.

PROBLEMS OF SUPPLY AND TACTICS

The flamethrower remained rare within the Waffen-SS during the early war years and despite the offensive credo fostered by the Waffen-SS, this organisation was slow in acquiring flame weapons. SS engineer units were not authorised flamethrowers prior to the onset of Operation 'Barbarossa'. Limitations of production and tactical applicability, as well as fears about the weapon's counter-use, all restricted German employment of the flamethrower. Moreover, its limited range meant that there was relatively little role for the weapon in the fast-paced offensive warfare the Germans practised early in the war. In addition, continued army hostility towards the Waffen-SS delayed acquisition of flame weapons. It was, therefore, only during the middle of the war that SS engineer companies received six flamethrowers. In fact, the flamethrower proved more valuable amid the positional defensive warfare the Germans primarily conducted in the last two years of the war. But by that stage the flamethrower was a luxury the Waffen-SS could ill afford, and their numbers dwindled during 1944–45 as productive capacity was diverted to more pressing needs.

In 1944 the personnel of the 500th and 600th SS Parachute Battalions received small quantities of a light, one-shot, disposable flamethrower, the Einstossflammmenwerfer 46, originally developed for the Luftwaffe. It was a self-contained cylindrical tube 597mm (23.5in) long with

Stielhandgrenate 39

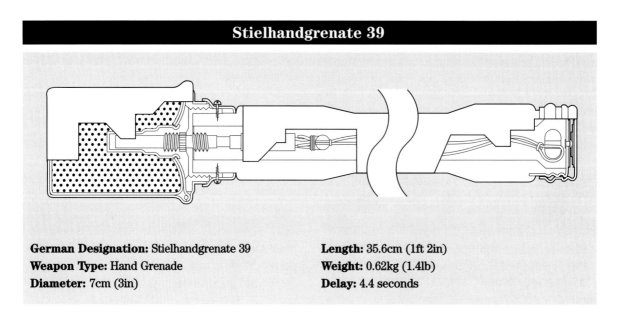

German Designation: Stielhandgrenate 39
Weapon Type: Hand Grenade
Diameter: 7cm (3in)

Length: 35.6cm (1ft 2in)
Weight: 0.62kg (1.4lb)
Delay: 4.4 seconds

fuel and propellant for a single burst, after which it was discarded. A simple trigger mechanism fitted at the forward end fired an explosive cartridge which provided the gas pressure to project a single burst of flame to a distance of about 27m (29.5yds) for half a second. This weapon was intended to be carried under the arm and fired like a pistol, but this left the user uncomfortably close to the flame as it left the stub nozzle! Most operators thus preferred to operate the weapon at arm's length, even though this entailed a resultant loss in accuracy!

FLAMETHROWERS AT ARNHEM

It was unusual, but not unheard of, for flamethrowers to serve in SS units other than engineer companies. Indeed, it was such a valuable offensive weapon that SS grenadiers long coveted the flamethrower as a means to reinforce their firepower. One such situation where this occurred was the commitment of the SS NCO School Arnhem in the defence of that town during the September 1944 Allied Operation 'Market Garden'. The school trained specially selected enlisted personnel for service as NCOs and put them through a vigorous and demanding NCO leadership course. On 17 September, it was business as usual at the school as it moulded another course of NCO candidates for future squad leadership. However, the drop of the 1st British Airborne Division in and around Arnhem provided a rude awakening. Alerted by the town commandant, the school commander quickly organised his 600 instructors and students into an alarm battalion and committed it to action. The battalion fielded eight flamethrowers that were no doubt on establishment to provide future squad leaders with knowledge of the proper offensive use of the weapon. The SS NCO aspirants used these flame weapons against the stubborn British resistance offered from the houses and cellars around Arnhem bridge. The British paratroopers had no defence against flame, which

proved highly effective in urban combat, penetrating houses that small arms could not. Flame bursts set many buildings alight and forced the defenders both to relinquish key buildings and move out into the open, where they were easily picked off. Such flame tactics played an important role in the successful German recapture of the Arnhem bridge over the Lower Rhine.

The SS Rock Climbing (Karstwehr) Battalion was another non-engineering unit that possessed six flamethrowers. This independent battalion, recruited during 1942, was directly subordinated to the Reichsführer-SS. It was tasked with internal security and anti-partisan operations in the inhospitable 'Karst' limestone mountains of the Balkans. Such terrain provided excellent refuge for communist guerrillas who were locked into a bitter and brutal war against the Axis occupiers. No doubt, the flamethrower was a terrible weapon against the irregular partisan forces the battalion engaged, and also one that buttressed the morale of the SS troopers, operating in an environment of hostility where potentially a partisan ambush awaited behind every corner. It was difficult enough for regular troops to hold in the face of flame attacks, let alone the admittedly brave but ill-trained irregulars of the anti-Nazi partisan movements. In late 1944 the SS High Command attempted to expand the battalion to divisional status, but the so-called 24th SS *Karstjäger* Division never exceeded reinforced regimental strength.

WAFFEN-SS EXPLOSIVES

Waffen-SS engineers also employed various demolition charges and anti-personnel mines to destroy enemy bunkers and bridges, to obstruct roads, and to immobilise equipment on the retreat. The most commonly used demolition charge in an offensive role was the Stangenladung, or pole charge, where explosive charges were attached to a pole and then thrust onto the apertures of enemy bunkers and then detonated.

Hollow-charge detonation charges were also used against steel and concrete obstacles. Mines were a defensive weapon employed in large numbers by all combatants of World War II, and Germany was one of their largest users. Mines essentially fall into two categories: anti-tank devices (which are discussed in Chapter 3), and the more common anti-personnel variant. The latter were much smaller and lighter and contained fragmentation explosive designed to inflict maximum injury on humans in the blast range.

MINE WARFARE

The purpose of mines is generally either to deny an area to an enemy or to slow down or channel enemy efforts to advance. Relatively unused in the early war years, the Germans began to employ mines in ever-increasing numbers from 1941, particularly on the thinly manned Eastern Front. Here mines proved a useful addition to defensive strength. In Western Europe, too, the Germans laid millions of mines as part of their Atlantic Wall defences and also sowed millions more along the Siegfried Line on Germany's western frontier. The first anti-personnel mine used by the Waffen-SS was the Schrapnellmine 35 (SMi 35), which the Allies termed the 'shoe mine'. It was buried with just igniters protruding above ground, or connected to trip wires. When tripped it ejected a small cylinder which scattered some 350 small steel balls over an area of 150m (164yds), sowing death and mutilation. The Stockmine operated on a similar principle to the SMi 35, but the device was fixed above ground on a wooden stake. The body comprised a concrete cylinder which held the charge and the shrapnel, which was scattered when a trip wire was moved. In 1943 the Germans introduced the Glasmine 43 (GLMi 43), which had a body made entirely of glass in order to reduce its detectability. Although inevitably part of the firing mechanism had to be metal, the mine was very difficult to detect. The Rollbombe was another simple

'mine' consisting of a concrete ball containing an explosive charge and shrapnel. The bomb was mounted on a slope and, when required, the igniter was pulled and the weapon literally rolled down the slope towards the enemy where it exploded. Such roll bombs were usually improvised in the field by engineers and therefore took on a multiplicity of forms, depending on what raw materials were available.

Hand grenades were first introduced into warfare during the mid-fifteenth century but until the late nineteenth century they were the preserve of specialised assault troops, known as grenadiers. It was the evolution of modern, industrial warfare in the late nineteenth century that saw the proliferation of hand grenades as they became an integral element of the infantryman's arsenal. They became so widely used during the trench warfare during World War I that they became as commonplace as the rifle.

SS HAND GRENADES

In September 1939, SS infantry went to war equipped with two basic types of hand grenades: the Stielgrenate stick grenade and the smaller Eiergrenate egg grenade. In the popular imagination the stick grenade has become synonymous with the German soldier of World War II, though both types saw widespread use. Both grenades had a thin metal casing and relied on the blast effect of the charge rather than on fragmentation of the casing. The Stick Grenade existed in two basic forms: the 1924 model and the longer 406mm (16in) 1939 design. Both worked on the same principle. Before throwing, the grenade was primed by pulling on the cord at the base of the handle. The soldier then had 4.5 seconds to throw the grenade overarm before it detonated. The advantage of the stick grenade was that the handle provided greater leverage, which allowed the grenade to be thrown farther and with greater accuracy. In fact, it was an SS trooper that held the World War II hand-grenade throwing record with a stick grenade throw of 70m (77yds)! For additional blast effect a special fragmentation sleeve could be attached to the top of the grenade to increase its shrapnel damage.

The ingenious Germans also found a way to use the stick grenade as a demolition charge in the guise of the Gebalteladung, which comprised six stick grenades minus their handles tied around a seventh complete stick grenade. So configured, they could even be used in an anti-tank role by being thrown under the tracks of approaching tanks. Devious SS troops also sometimes replaced the delay detonator with an instant detonator in grenades 'apparently' left behind in an ostensively hurried withdrawal in the hope that the grenades would be picked up and used by enemy troops. The various egg grenades used by the SS were smaller and less powerful than the stick grenades. The Eierhandgrenate 39 used a thin metal case, and the thrower unscrewed a protective cover to prime the igniter. The grenade contained 0.35kg (0.75lb) of explosive and had a blast radius of about 13m (14.2yds). The one tactical advantage of the egg grenade, though, was that it could be thrown either over or underarm.

During 1941, SS-Sturmbannführer (Major) Kurt 'Panzer' Meyer became renowned within the German military for devising a unique employment of hand grenades to motivate his troops. Meyer commanded the reconnaissance battalion of the *Leibstandarte* Division during the Balkan Campaign in the spring of 1941. His battalion was held up at the Klissura Pass, which was strongly held by Greek troops. When murderous Greek defensive fire pinned down Meyer's battalion as it attempted to force the pass, its commander hit on a novel way of maintaining the advance: he dropped a primed egg

grenade behind the last man in each section! Faced with the prospect of being blown up, the SS troops stormed through the hail of Greek fire and rapidly took the stubborn Greek positions in bloody hand-to-hand combat!

RIGHT: *A soldier of the* **Das Reich** *Division in the Balkans, spring 1941. He carries a Stielhandgrenate 39 hand grenade tucked into his belt.*

CHAPTER 3

Anti-tank and Anti-aircraft Weapons

Integral to the martial prowess of the Waffen-SS was its ability to counter enemy armour on the battlefield, and hostile aircraft in the air.

The Waffen-SS employed a variety of anti-tank weapons during World War II to counter the increasing battlefield threat posed by Allied armour. These weapons included anti-tank rifles, anti-tank guns, anti-tank rockets and anti-tank mines, all described in this chapter. The SS also employed self-propelled anti-tank guns (Panzerjäger) and tank destroyers (Jagdpanzer) which are discussed in Chapter 7. German anti-tank weapons proved an economical and effective counter to the firepower and mobility of enemy armour, and they provided defensive staying power as SS troops found themselves confronted by ever-greater numbers of enemy tanks.

LEFT: *An SS 7.5cm Pak 40 anti-tank gun engages the advancing enemy on the Eastern Front, the smoke on the horizon indicating the ferocity of the struggle.*

However, the offensive orientation of pre-war German doctrine, procurement and training ensured that the Waffen-SS remained deficient in anti-tank weaponry during the early war years. It was this inadequacy that provided the impetus for German research that ultimately produced some of the most lethal and novel anti-tank weapons of World War II.

Germany was the first country to develop and use the anti-tank rifle during World War I. The Germans designed the 13mm Mauser 'Tank Gewehr' rifle that fired a solid-steel shot. During the 1920s, however, the anti-tank rifle fell into disfavour among the German military, and it was not until the Nazi rearmament of the late 1930s that Rheinmetall introduced the 7.92mm Panzerbüsch (PzB) 38 anti-tank rifle. It could penetrate 25mm (1in) of 30-degree sloped armour at 300m

(328yds). However, it was a complex weapon with a sophisticated recoil mechanism in which the barrel actually recoiled within the stock, designed to save the gunner's collar-bone from dislocating on firing. The rifle weighed 15.9kg (35lb) and was of conventional design with pistol grip, padded shoulder stock and bipod mount. The recoil mechanism was soon found to be a waste of resources and production ceased after only 1600 had entered service.

While the SS received small numbers of this weapon during 1939 – there were only 538 operational at the outbreak of the war – an improved successor, the 7.92mm PzB 39, soon superseded it in service. This anti-tank rifle was much simpler since a muzzle brake replaced the sliding-barrel mechanism. Overall weight therefore declined to 12.35kg (27.2lb) and performance was slightly better. Manufactured in large numbers – no less than 25,298 were in service by June 1941 – the rifle served in the infantry regiments and anti-tank battalions of the *SS-Verfügungs* and *Totenkopf* Divisions in the May 1940 Western

BELOW: *A 7.92mm Panzerbüsche 38 anti-tank rifle. Note the bipod mount, shoulder stock and taper-bore barrel for increased muzzle velocity.*

Campaign, where it proved barely able to hold its own. By the summer of 1941, better armoured tanks had rendered it obsolete.

STOP-GAPS NOT ENOUGH

To extend its service life, the Germans introduced a new tungsten-core bullet with greater hitting power in 1940. Tungsten is a very dense metal that increased the penetration of the of the PzB 39 to 33mm (1.3in) at 300m (328yds). This stop-gap measure could only temporarily prolong the service life of the weapon, however, and after 1941 the anti-tank rifle was relegated to second-line formations and training units, where it continued to serve until the end of the war. The 12th SS Panzer Division *Hitlerjugen*d, however, retained 72 PzB 39 as late as the summer of 1943.

It was only late in the war, as Germany's plight became ever more desperate, that anti-tank rifles once again saw occasional frontline action as the SS threw every available weapon, however obsolete, into battle. One such case was the commitment of the SS NCO School based at Arnhem in an attempt to thwart the September 1944 Allied Operation 'Market Garden'. The school deployed six obsolete PzB 39 anti-tank rifles along the Waal River in an effort to block

Panzerbüsche 39

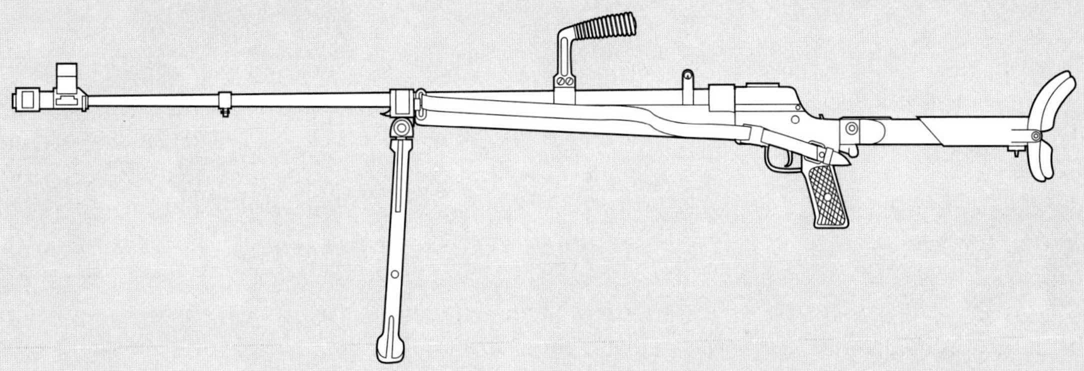

German Designation: 7.92mm Panzerbüsche (PzB) 39

Weapon Type: Anti-tank Rifle

Calibre: 13-7.92mm (0.51-0.312in) Taper Bore

Length: 2.7m (8ft 10in)

Weight: 12.6kg (27.8lb)

Muzzle Velocity: 1140m/s (3738ft/s)

Armour Penetration: 25mm (1in) at 300m (328 yds) – (30 degrees)

the Allied advance on the Reich. The 16th SS Panzergrenadier Training and Replacement Battalion, also headquartered at Arnhem, fielded several more of these aged anti-tank rifles. In the aftermath of the drop of the 1st British Airborne Division on 17 September, these weapons proved not altogether ineffective against the lightly armoured vehicles glider-landed in support of the British drop.

The next logical step in the race to find an effective counter to the tank was to develop a specialised anti-tank gun: a relatively low-calibre, mobile weapon capable of firing solid shot at high velocity to penetrate the armour of enemy tanks. The first viable German anti-tank gun, the 3.7cm Pak L/45, emerged in 1928. It was a small, mobile horse-drawn weapon and for its day it represented an excellent design. With a low-silhouetted, light and easy to conceal, it could penetrate the armour of almost any tank then in service. Redesigned for towing by motor vehicles, the modified gun re-emerged in 1935 as

the 3.7cm Pak 35/36. It became the standard anti-tank gun of the German Army and Waffen-SS early in the war and was produced in very large numbers, in excess of 20,000. The gun first saw action in 1936 with the German Condor Legion during the Spanish Civil War. The gun weighed only 432kg (952.5lb) and had a sloping splinter shield. The barrel was some 42 calibres in length and fired a solid-shot round at a muzzle velocity of 762m/s (2500ft/s) to a maximum range of 4025m (4400yds), though obviously its chances of success at such extreme range were very slim. At a range of 500m (547yds) the gun could easily penetrate 48mm (1.9in) of vertical homogenous armour or 36mm (1.4in) of 30-degree sloped armour.

The 3.7cm Pak 35/36 became the first anti-tank gun weapon to serve with the Waffen-SS. At the start of World War II in September 1939, the four armed SS motorised infantry regiments – *Leibstandarte, Deutschland, Germania,* and *Der Führer* – each possessed a towed regimental

ABOVE: *A 3.7cm Pak 35/36 anti-tank gun of the* SS-Verfügungs *Division (later the* Das Reich *Division) in the West in the summer of 1940.*

anti-tank company equipped with 12 of these guns. In addition, the independent SS Reconnaissance Battalion had a towed anti-tank platoon with three additional guns. In total, therefore, at the start of the war the SS deployed in the field a mere 51 anti-tank guns. When one considers that in September 1939 the German Army fielded 11,200 Pak 35/36 guns, one realises how fledgling the Waffen-SS remained at the start of the war.

It was in Poland during September 1939 that the SS gained its first, if albeit very limited, experience of anti-tank warfare. In Poland the Pak 35/36 proved more than adequate for operational needs in the face of relatively modest armoured opposition. The Poles had little medi-

um armour and they dispersed their more numerous light tanks and tankettes for infantry support and to cavalry units. After the Polish Campaign, the Waffen-SS raised two additional motorised anti-tank battalions for both the *SS-Verfügungs* Division and the newly formed *Totenkopf* Division. These battalions possessed the conventional organisation of the period: three companies of 12 3.7cm Pak 35/36s. In addition, the *Totenkopf* motorised reconnaissance battalion formed a towed anti-tank platoon equipped with a further three guns. Thus as the start of the Campaign in the West on 10 May 1940, the number of Pak 35/36 guns in frontline SS service had risen to 90.

It was in France in 1940 that the German Army first came to appreciate the tactical limitations of the Pak 35/36 as German forces encountered Allied heavy tanks – the French Char B and the British Matilda – whose thick armour proved all

but impervious to 3.7cm anti-tank rounds. It was the disastrous encounter between the *Totenkopf* Division and British Matildas near Arras on 20 May 1940 that exposed the inadequacy of the 37mm anti-tank gun. At Arras, weak Anglo-French forces counterattacked the exposed flank of the advancing *Totenkopf* Division and Erwin Rommel's 7th Panzer Division. Though the Matilda was armed with only two Vickers 0.5in machine guns and could not exceed a slothful 13kph (8mph), the tank possessed armour up to 60mm (2.4in) thick. Two Matildas ran amok through several hastily deployed SS defensive positions and could not be halted, overrunning two *Totenkopf* anti-tank companies and knocking out four of Rommel's panzers. To the consternation of the SS gunners, their 3.7cm shells simply pinged off the Matilda's thick armour, even at point-blank range. German examination of one

BELOW *A 7.5cm Pak 38/97(f) – the barrel of a former French 75mm mle 1897 field gun married to the 5cm Pak 38 carriage – in action in Normandy in 1944.*

of the Matildas after it had been eventually disabled revealed that it had shrugged off no less than fourteen 3.7cm hits. Stunned by the ineffectiveness of their guns, the inexperienced *Totenkopf* anti-tank crews abandoned them when they came under fire and bolted to the rear, taking with them a number of SS infantry companies and an entire artillery battery in a disorderly rout back through Wailly. Fortunately for the panicked SS troopers, Erwin Rommel was close at hand, leading as usual from the front. Rommel quickly organised a defence in depth utilising all available anti-tank guns as well as field guns, howitzers, and borrowed Luftwaffe 8.8cm flak guns that fired over open sights.

The first SS encounter with Allied heavy armour was thus a very ignominious one. The

BELOW: *A frontal view of a 7.5cm Pak 40 heavy anti-tank gun. Its large splinter shield and low silhouette are clearly evident in this photograph.*

Totenkopf was a young and green division, its personnel drawn from concentration camp guard units and lacking an experienced officer and NCO corps. As a result, elements of the division panicked and hastily withdrew in disorder. The British inflicted 700 casualties, including 400 prisoners – the majority from the *Totenkopf* – and knocked out 20 German tanks and a score of anti-tank guns. Though the British attack force was meagre and its achievements limited, the battle exerted a significant psychological impact on the Germans, who were unused to serious counterattacks. The *Totenkopf*'s disorderly flight at Arras became a blot on the military reputation of the Waffen-SS.

German forces also experienced similar outcomes in engagements with the French Char B tank, which proved just as impervious to German 3.7cm anti-tank fire. Consequently, the German military concluded that the army needed a heavier anti-tank gun. The bitter struggle on the

Pak 40

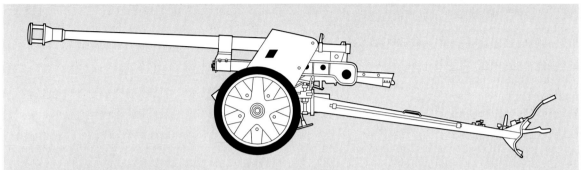

German Designation: 7.5cm Panzerabwehrkanone
  (PaK) 40
Weapon Type: Heavy Anti-tank Gun
Calibre: 7.5cm (2.95in)
Length (of piece): 3.7m (12ft 2in)
Weight (in action): 1425kg (3142lb) = 1.425
  tonnes (1.42 tons)
Traverse: 65 degrees

Elevation: -5 to +22 degrees
Muzzle Velocity: PzGr 39 = 750m/s (2459ft/s);
  PzGr 40= 933m/s (3059ft/s)
Maximum Range: AT rnd = 2000m (2188yds);
  HE = 7680m (8402yds)
Armour Penetration: 94mm (3.7in) at 1000m
  (1094yds) – (30 degrees)

Eastern Front after June 1941 painfully reconfirmed the growing obsolescence of the Pak 35/36, which was deployed in large numbers: 14,459 were in service at the start of Operation 'Barbarossa'. The gun proved so ineffective against the new Soviet T-34 and KV-1 tanks that its crews derisively dubbed it the 'door-knocker'! This was the sentiment of *Totenkopf* anti-tank gunners at Lushno on the Pola River on 24 September 1941, when the division encountered its most serious Soviet armoured counterattack to date with 20 T-34 tanks engaging them. Rounds from the 3.7cm Pak 35/36 simply pinged off the frontal armour of the T-34 and the divisional artillery had to save the day, firing over open sights to halt the attack that had penetrated deep into the *Totenkopf*'s defences.

The introduction of two new rounds that enhanced performance temporarily extended the service life of the Pak 35/36. The first of these was the 3.7cm Panzergrenate 40 (PzGr 40) tung-

sten-carbide round introduced in late 1940, which could penetrate 65mm (2.6in) of vertical armour and 55mm (2.2in) of 30-degree sloped armour at 500m (547yds). However, growing shortages of tungsten due to the Allied naval blockade ensured that the round could only be produced in limited quantities. Superior still was the 3.7cm Stielgrenate 41 stick bomb introduced in the autumn of 1941. The bomb fitted over the muzzle of the gun and was fired by means of a blank cartridge in the breech, in exactly the same fashion as a rifle fired a rifle grenade. The bomb was fin-stabilised and contained a hollow-charge HEAT (High Explosive Anti-Tank) round for use against armour.

In a hollow-charge round, a cone-shaped explosive charge detonates forward upon impact, creating a pressure wave of molten plasma which blasts a tiny hole through armour plate allowing the high-explosive warhead to penetrate and detonate inside a tank, killing or maiming the crew

ABOVE: *A Waffen-SS Pak 40 in action on the Eastern Front. Note the 'kill' rings on the barrel – testament to this weapon's effectiveness against enemy tanks.*

within. Though it possessed a range of just 300m (328yds), the Stielgrenate 39 increased the penetration of the Pak 35/36 to 180mm (7.2in), more than sufficient to destroy any tank then in service. The stick bomb thus extended the operational life of the Pak 35/36 with the Waffen-SS until more powerful anti-tank weapons could supplant it at the front. In the case of the *Totenkopf* Division, it was not until August 1942 that this could be accomplished and in the meantime the stick bomb provided valuable service.

As heavier anti-tank guns entered service during 1941–43, the Waffen-SS increasingly relegated its remaining Pak 35/36 to foreign volunteer, training and security units, where they continued to serve in diminishing numbers until the end of the war. In June 1944, for example, the SS garrison forces deployed in the Netherlands fielded 35 Pak 35/36 anti-tank guns.

Long before Arras revealed the deficiencies of the 3.7cm Pak 35/36, Rheinmetall-Borsig had begun working on a larger gun, the 5cm Pak 38, which was destined to become the most widely used SS anti-tank gun of the war. Like its predecessor, the gun was light and mounted on a light, split-trail carriage that was easy to manhandle.

The first two guns entered service in May 1940 and full-scale production began during the latter half of 1940. During the spring of 1941 the first guns began to equip a single platoon in the anti-tank companies of select army and SS divisions. In service it soon proved superior to the Pak 35/36. Weighing 986kg (2174lb), it possessed rubber-tyred wheels, a torsion-bar sprung carriage and a sloped double splinter shield. The gun fired a variety of ammunition: its basic Panzergrenate 38 round had a maximum range of 2650m (2900yds) and could penetrate 61mm (2.4in) of vertical and 50mm (2in) of 30-degree sloped armour at 1000m (1094yds), just enough to do some real damage to a Soviet T-34 tank. Late in 1940 the Germans introduced a tungsten-carbide round for the gun, which increased its performance sufficiently to penetrate the frontal armour of the Soviet T-34 tank at typical combat ranges.

THE MIGHTY PAK 38

Combat affirmed the faith that the German High Command placed in the new anti-tank gun. One such action involved troops of the *Totenkopf* Division, whose courage and determination made their formation a legend. On 24 September 1941, at Lushno on the Pola River, SS-Sturmmann (Corporal) Fritz Christen saw every member of his platoon killed during a large Soviet assault supported by T-34 and KV-2 tanks. Despite being wounded and out of food and water, Christen remained at his post and single-handedly manned his 5cm Pak 38, repulsing all Soviet attacks for two days until help arrived, knocking out 13 Soviet tanks in the process. For his astonishing feat of heroism, Hitler personally decorated Christen with the Knight's Cross, the first and youngest Waffen-SS enlistee to receive this coveted award.

Even as the 5cm Pak 38 encountered Soviet heavy armour for the first time during the late summer of 1941, German designers realised that a more powerful gun would soon be needed.

Preliminary work on a new 7.5cm anti-tank gun had actually begun in 1939 and the first 7.5cm Pak 40, as the gun was called, reached army and SS troops on the Eastern Front in late 1941. The gun immediately proved its worth in combat and remained in production until the end of the war. It proved hard-hitting, easy to conceal and relatively cheap to produce. Moreover, the gun was desperately needed over the winter of 1941–42 as the Germans found themselves thinly stretched across the Eastern Front against an undefeated enemy. Its effectiveness ensured that troops clamoured for the weapon, but there were never enough Pak 40 guns to meet demand.

HEROISM OF THE *HITLERJUGEND*

Typical of so many tank-versus-anti-tank engagements was that of a platoon of the 12th SS Anti-Tank Battalion near St. Sylvain, Normandy, on 8 August 1944. This was the opening day of the massive Canadian offensive, Operation 'Totalize', intended to capture Falaise and complete the encirclement of the bulk of the German forces fighting in Normandy. Near the Chateau du Fosse, the 12th SS Divisional Escort Company, reinforced by two expertly camouflaged SS 7.5cm Pak 40 anti-tank guns deployed at the edge of a wood, opposed the advance of the 1st Polish Armoured Regiment across the open cornfields that bordered the woodland. In quick succession, the two guns knocked out six tanks before the Polish armour was able to return fire. The Polish tanks penetrated the thin SS defensive positions and the anti-tank guns destroyed another pair of Sherman tanks right in front of the company command post, before the Polish armour retired behind the Chateau and called down artillery fire support. This curtain of fire destroyed one of the anti-tank guns and neutralised the crew of the other. Seizing their opportunity the Polish armour resumed their advance, but they had not reckoned with SS grenadiers Kurt Breitmoser and Leo Freund, who manned the surviving gun. These impromptu gunners destroyed a Polish Sherman that was almost upon them with their first shot. But after another round, their position was hit, Breitmoser was knocked unconscious and Freund wounded. When Freund returned after receiving hasty medical attention nearby, he found his comrade had regained consciousness and together the pair resumed manning the gun. Disheartened by such stiff opposition, the Polish armour withdrew, having lost 22 tanks in the battle.

MEASURES TO MAKE UP NUMBERS

But the Pak 40 remained in very short supply, especially during 1942, and as a stop-gap the Germans converted some 700 French 1897 model 7.5cm field guns captured in the West during 1940 into improvised anti-tank guns, mounting them on existing 5cm Pak 38 and 7.5cm Pak 40 carriages. The Germans rushed the resulting weapons, designated the 7.5cm Pak 97/38 and 97/40, to the Eastern Front. The hasty improvisation had its drawbacks, however, the most serious of which being that the weapon was really too powerful and became unstable on firing. As the 7.5cm Pak 40 became more readily available during 1943, the SS relegated its remaining 7.5cm Pak 97/38 and 97/40 guns to training and rear-area security units, where they continued to serve until the end of the war.

Another anti-tank gun rushed into SS service to compensate for the limited availability of the Pak 40 was the modified 1936 model Soviet field gun, which was captured in significant numbers during Operation 'Barbarossa'. The Germans rechambered the gun to take the standard 7.5cm PzGr 40 anti-tank round and many also had a muzzle-brake fitted. This modified weapon, designated the 7.62cm Pak 36(r), soon became regarded as one of the best anti-tank guns of the war, despite its hastily improvised nature. It possessed good penetration, was mobile and proved reliable in service. Smaller numbers of the Soviet

1939 model field gun underwent similar conversion as the 7.62cm Pak 39(r). The performance of the two guns was only slightly inferior to that of the Pak 40. Firing the PzGr 40 round, the Pak 36(r) could penetrate 84mm (3.3in) of unsloped armour at 2000m (2188yds). Against 30-degree sloped armour at the same distance the respective penetration was 55mm (2.2in). Both guns were of conventional design, with split-trial carriage, rubber-tyred wheels and bolted-on splinter shield. They did, however, have high profiles for anti-tank guns, which made them more vulnerable than the 7.5cm Pak 40.

Waffen-SS gunners demonstrated the efficacy of the weapon in a counterattack by the 3rd Battalion, 26th SS Panzergrenadier Regiment against the bridgehead established over the Orne River by the 59th British Infantry Division and the 107th Regiment, Royal Armoured Corps, at Grimbosq, Normandy, on 7 August 1944. Two 7.62cm Pak 36(r) from the 1st Battalion of the same regiment deployed at the edge of the Grimbosq forest supported the furious SS counterattack. Between them the two guns knocked out no less than 24 British tanks and proved the decisive element in the recapture of the hamlets of Grimbosq and Brieux.

ONE MAN AND HIS PAK 40

During 1942–44, the 7.5cm Pak 40 gradually supplanted the 5cm Pak 38 and the improvised 7.5cm Pak 97 and 7.62cm Pak 36(r). In service the Pak 40 proved a powerful and deadly weapon in the hands of trained anti-tank gunners, and the gun became the most prolific German anti-tank weapon of the war. A young SS-Sturmmann (private), Remi Schrijnen, of the SS Sturmbrigade *Langemarck* demonstrated this at a strategic position, grimly dubbed by its SS defenders as 'Orphanage Hill'. This position comprised part of the Tannenberg Line defences behind the River Narva on the northern sector of the Eastern Front. Here, on 2 January 1944, Schrijnen and his

7.5cm Pak 40 turned back a large Soviet armoured force, destroying three Red Army tanks in the process. On 3 March 1944 he was wounded for the seventh time, but refused hospitalisation and remained with his unit. Three days later the convalescing Schrijnen, after witnessing his entire crew being killed, single-handedly manned his gun in disobedience of orders to retire and continued an apparently hopeless struggle against 30 Soviet tanks, many of which were the latest Josef Stalin (JS)-II heavy tank. Schrijnen destroyed 11 Soviet tanks, including three JS-II vehicles, before his gun was destroyed and he was seriously injured. A German counterattack subsequently found him battered but alive, and he was deservedly awarded the Knight's Cross for his exceptional heroism.

But the Germans realised that eventually an even heavier gun would be needed given the rate of increasing armour on tanks. During 1943 Krupp designed the 8.8cm Pak 43, the finest anti-tank gun to see operational service during World War II. Mounted on a low, easy-to-conceal cruciform platform, the Pak 43 was hard to spot and easy to handle. More importantly, it could knock out any tank then in service. It arrived at the front in numbers for the first time in 1944, just in time to counter the next generation of heavy Soviet armour, the JS-I and JS-II, which were almost invulnerable to the Pak 40. As a dedicated anti-tank gun it proved far superior to the improvised dual-purpose 8.8cm Flak 41. The Pak 43 had a long 71-calibre barrel with muzzle brake and its low-slung carriage reduced its height to only 2.05m (6ft 9in). A well-sloped splinter shield further reduced its silhouette and hence its vulnerability. The gun could even be fired from its firing platform while limbered, though over a restricted traverse, a very useful tactical capability. At 3700kg (8160lb) it also weighed much less than the Flak 88. The gun had an advanced semi-automatic breech loader and electrical firing circuit that allowed rapid fire of a new 10.4kg (22.9lb) armoured-piercing round. This round

Pak 43

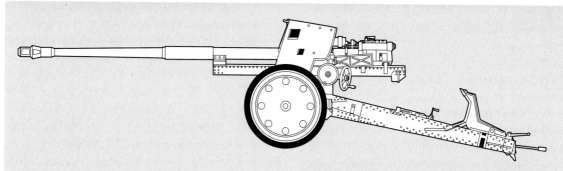

German Designation: 8.8cm Panzerabwehrkanone (Pak) 43
Weapon Type: Heavy Anti-tank Gun
Calibre: 8.8cm (3.5in)
Length (of piece): 6.61m (21ft 8in)
Weight (in action): 3650kg (8048 lbs) = 3.65 tonnes (3.6 tons)

Muzzle Velocity: PzGr 39/43 rnd = 1000m/s (3279 ft/s); 40/43 = 1130m/s (3705 ft/s)
Maximum Range: HE = 15,150m (16,574 yds)
Armour Penetration: PzGr 39/43 = 167mm (6.6in) at 1000m (1094 yds) – (30 degrees); 40/43 = 192mm (7.6in)

gave phenomenal penetration: 207mm (8.1in) against unsloped and 182mm (7.1in) against 30-degree sloped armour at a range of 500m (540yds). Even at 2000m (2188yds) the gun could penetrate 159mm (6.3in) and 123mm (4.85in) of vertical and 30-degree sloped armour respectively. With an impressive range of up to 4000m (4376yds), the Pak 43 was capable of defeating any Allied tank in service in 1943.

Despite its simplicity, the Pak 43 was neither easy nor cheap to produce as it required high-quality materials that were increasingly in short supply. As a result, production was limited and demand for the weapon far exceeded supply. It remained the rarest anti-tank gun in SS service and only a few formations ever received the gun. None had reached the SS formations stationed in France prior to the D-Day landings on 6 June 1944. What the gun was capable of, however, was demonstrated on 17 July 1944, southeast of Caen, during General Bernard Montgomery's Operation 'Goodwood' offensive. Buttressing the defence of I SS Panzer Corps were 54 8.8cm Pak 43 guns of the 1039th and 1053rd Army Heavy Motorised Anti-tank Battalions. Though the massive Allied air offensive that preceded the offensive destroyed 36 of these guns, three intact 8.8cm Pak 43 guns deployed in woods east of Cagny knocked out 23 tanks of the 11th British Armoured Division and held up the entire advance for hours.

THE FEARSOME 8.8CM PAK 43

It was only late in 1944 that the 8.8cm Pak 43 began to reach SS formations. The 17th SS Panzergrenadier Division *Götz von Berlichingen*, for example, received nine guns over the winter of 1944–45 and retained two of them as of 2 February 1945. The 18th SS Panzergrenadier Division *Horst Wessel* likewise received a company of nine guns during the spring of 1945, one of which remained operational as late as 1 April 1945.

By the spring of 1945, a combination of mounting losses, dire shortages of towing trac-

ABOVE: *The first version of the Panzerfaust anti-tank weapon entered service in 1942. This weapon gave Waffen-SS grenadiers a viable anti-tank capability.*

tors and petroleum, as well as increasing diversion of production resources to tank destroyers and assault guns, saw the number of anti-tank guns in Waffen-SS service plummet. A strength return of 1 April 1945 revealed that there were barely 100 7.5cm Pak 40 anti-tank guns left in Waffen-SS service at this late stage of the war.

Inevitably the SS acquired limited quantities of a variety of captured foreign anti-tank guns to supplement its meagre stock of German weapons. Among the earliest foreign anti-tank guns to see limited service in the Waffen-SS were the 2.5cm Pak 112(f) and 113(f). The standard French light anti-tank guns of 1940, they were captured in some numbers during the invasion of the West. Though light and mobile, the guns had short range, only 1800m (1971yds), and poor penetration: a mere 50mm (2in) at 500m (547yds). Consequently, the Waffen-SS quickly relegated them to training and garrison forces, though the 17th SS Panzergrenadier Division *Götz von Berlichingen* retained three guns as late as the spring of 1944. A more common weapon was the 3.7cm Pak 37(t). The standard Czech anti-tank gun of 1938, it was an efficient design with similar performance to the Pak

35/36. It saw service with the *Totenkopf* and *SS-Polizei* Divisions early in the war.

Among medium anti-tank guns, the Waffen-SS eagerly sought to acquire the 1936 model Skoda 47mm gun, which in German service was designated the 4.7cm Pak 36(t). The gun could penetrate 60mm (2.4in) of homogenous armour plate at 1200m (1314yds). After 1941 the remaining guns were relegated to garrison units. Another medium anti-tank gun that saw limited service in the Waffen-SS was the French 4.7cm gun. Designated the 4.7cm Pak 181(f) in German service, it served with SS garrison forces in Western Europe.

Quantities of Red Army anti-tank guns also fell into SS hands. These included the 4.5cm Pak 184(r) and 5.7cm Pak 208(r) medium guns. Likewise, after Italy's capitulation in September 1943

*BELOW: **Two Waffen-SS soldiers wait for the next Red Army attack on the Eastern Front. A Panzerfaust rests between them on the lip of their trench.***

a variety of Italian anti-tank weapons found their way to SS units deployed in Italy and the Balkans. The former Italian anti-tank gun that saw the most widespread SS service was designated the 4.7cm Pak (Böhler) 177(i). Though it lacked hitting power, it continued in SS service after the German occupation of Italy. The 29th SS Grenadier Division (Italian No 1), for example, received two companies of the gun during the spring of 1945. It is unclear whether any ever saw action, however, since the division never completed its activation before the German collapse in April 1945.

Despite ever-increasing production and the utilisation of captured weapons, there were never enough anti-tank guns to protect every German infantry unit. What was needed, the Germans ultimately realised, was a light, portable infantry anti-tank weapon that could be mass produced. German experiments with hollow-charge grenades appeared to point the way

forward, since they could be fired from a relatively simple low-velocity weapon. During 1942–43, the Germans developed revolutionary new personal hollow-charge infantry anti-tank weapons. Though technically infantry support weapons, these anti-tank devices are included here for reasons of clarity and completeness.

THE 'ARMOURED FIST'

Hollow charges allowed the development of anti-tank warheads of much smaller size. The result was a brand new type of discardable, single-shot anti-tank rocket, dubbed the Panzerfaust – literally 'Armoured Fist' – though the weapon was officially called the Faustpatrone ('Fist Cartridge'). The first model, the Faustpatrone 30, entered service in 1942. It was a very simple device that consisted of a hollow steel tube 36cm (14in) long with a hollow-charge grenade fitted at one end and an explosive charge in the centre providing a counter-blast that projected to the rear to eliminate recoil, in similar fashion to a recoilless gun. The earliest version had a very powerful back-blast that required the weapon to be held at arm's length, which made it very difficult to aim and nearby troops had to be careful not to be caught in the blast. Later production models had an elongated tube that allowed the weapon to be tucked under the arm and fired, which greatly improved accuracy. Weighing only 5kg (11.25lb), its hollow-charge warhead could penetrate 140mm (5.5in) of armour plate sloped at 30 degrees, quite sufficient to disable both the T-34 and the KV-1. The grenade was stabilised in flight by fins that folded onto the shaft of the grenade as it was loaded and then sprang out once in flight. The biggest drawback of the weapon was its extremely limited range of only 30m (66ft). The earliest Panzerfaust anti-tank weapons, therefore, had limited tactical utility. Nevertheless, the basic design proved sound and suggested possibilities for the future.

During 1943 the Germans developed an improved weapon of extended 60m (67yds)

range, the Faustpatrone 60. In turn this was replaced in 1944 by the Faustpatrone 100, which had a double propellant charge to increase its range to 100m (109yds). During the last year of the war the Panzerfaust was produced in huge quantities and widely issued to infantry tank-destruction units. The Panzerfaust finally provided SS grenadiers with the anti-tank capability for which they had long clamoured. During the last two years of the war, enemy armour thus became much more vulnerable to German infantry whenever they operated without infantry support.

PANZERSCHRECK – 'TANK TERROR'

Alongside these rather short-range weapons the Germans developed one of the best infantry anti-tank weapons of World War II, the Raketenpanzerbüsche (RPzB) 54 anti-tank rocket-launcher. This weapon was based upon the American Bazooka, which German troops first encountered in Tunisia. It so impressed the Germans that they rapidly produced a superior copy. The weapon, named the Panzerschreck ('Tank Terror'), but more colloquially (and derisively) dubbed the Ofenrohr ('Stovepipe'), consisted of a metal tube that fired a rocket-propelled hollow-charge anti-tank warhead. The tube came with a very basic shoulder stock and pistol grip and was constructed of simple, lightweight materials throughout. A small, rounded shield was mounted in front of the trigger mechanism and contained a small round transparent plate on the left-hand side to allow sighting while the shield protected the operator from the rocket's powerful back-blast produced as the rocket grenade exited the tube. The fin-stabilised projectile weighed 3.3kg (7.2lb) and left the tube at a velocity of 105m/s (346ft/s). Unlike the Panzerfaust, however, the weapon was reusable and was, therefore, not discarded after firing. The unloaded weapon weighed 9.2kg (20.3lb) and possessed an effective range of 100m (109yds). In service it proved a more accurate

weapon than the Panzerfaust and had an additional advantage in that it could be fired from a prone, rather than from a kneeling or standing, position.

During the last year of the war, German forces came to rely increasingly on these tank-destruction weapons to provide anti-tank defence against the ever-increasing numbers of Allied tanks. However, successful close-combat tank destruction required iron nerves and fanatical spirit. Consequently, the High Command often employed SS anti-tank units in desperate defensive battles against greatly superior numbers of enemy infantry and armour. One such unit was the SS Tank Destruction Company *Dora II*. Formed from Otto Skorzeny's SS Commandos, *Dora II* was equipped with Panzerfausts, Panzerschrecks, anti-

BELOW: *The Panzerschreck was based on the American Bazooka. Note the shield, which protected the firer from the rocket's powerful back-blast.*

tank mines and light infantry weapons. During April 1945, it deployed behind the Seelow Heights east of Berlin with the mission of preventing at any cost an operational breakthrough by Soviet armour towards Berlin. The company was led by SS-Untersturmführer (2nd Lieutenant) Porsch, a much-decorated 20 year-old SS 'veteran' who had earned an Iron Cross First Class, the coveted German Cross in Gold, and an impressive four awards of the tank-destruction badge on the Eastern Front. Skorzeny selected Porsch for command due to his experience as a tank hunter and his fanatical commitment to National Socialism.

Flak 30

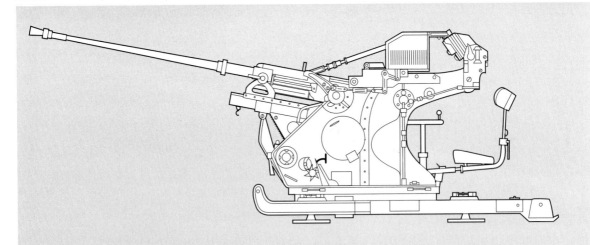

German Designation: 2cm Flugabwehrkanone
(Flak) 30
Weapon Type: Light Anti-aircraft Gun
Calibre: 2cm (0.8in)
Length: 2.3m (7ft 7in)
Weight: 463kg (1020lb)

Traverse: 360 degrees
Elevation: -10 to +100 degrees
Muzzle Velocity: 900m/s (2951ft/s)
Maximum Ceiling: 2200m (2407yds)
Rate of Fire: 120-280rpm

After the Soviet breakthrough onto the Seelow Heights of 16–19 April 1945, *Dora II* found itself among a sea of Soviet armour that poured through the gaps in the shattered German frontline and recklessly pushed forward. The calm Porsch chose his ground and established a well-organised defensive hedgehog around the town of Seelow, which commanded the main road to Berlin. The Soviets confidently sent in a regiment of heavy armour to sweep aside the SS resistance and open the road to Berlin. But Porsch and his company stood firm and repulsed the attack, leaving six burning Soviet tanks on the battlefield. The Soviets then sent an infantry regiment into the attack but that too was repulsed, despite its overwhelming numerical advantage. Having won for itself a brief respite, Porsch pulled back *Dora II* to Lebus in an effort to relieve the garrison, which was besieged by the Red Army. Though it arrived

too late to prevent the fall of the town, that day the company scored its 100th – Porsch his 13th – kill. Days of bitter defensive fighting followed. On 20 April, the company claimed its 125th kill and Porsch his 17th during a daring counter-attack that routed a Soviet rifle battalion at Neu Zittau and captured its headquarters staff. For this daring and successful counter-strike, the German High Command awarded Porsch the Knight's Cross on 26 April 1945, one of the last to be awarded during the war. That same day in another counterattack, *Dora II* silenced a Soviet mortar battery and captured eight enemy mortars, which were quickly pressed into use against their former owners by the SS troopers.

His counter-thrust, however, put Porsch's unit into a salient which Soviet forces advancing on his flanks nipped off the following day, forming a pocket cut off behind Soviet lines. Nevertheless, *Dora II* continued to offer ferocious resistance against the Soviet efforts to subdue the pocket. Its defiance attracted to the unit those Wehrmacht stragglers who had been isolated from their units by the Soviet breakout but who nonetheless still possessed some fight in them. The group was also joined by civilians – women, children and old men – who desperately sought to escape the brutality that the Red Army desired to inflict on their hated enemy in revenge for the widespread appalling atrocities that the Nazis – including Waffen-SS forces – had visited upon the Soviet people. Porsche tried to fight his way through the Soviet lines, but the enemy strength grew until a breakout became impossible.

Between Markisch-Buchholz and Töpchin the last act was played out. Surrounded by a Soviet rifle battalion the SS troopers scorned capitulation, though most of the army troops accepted a

Soviet demand for honourable surrender offered out of respect and admiration for the fanatical resistance that the defenders had offered. Faced with no alternative but death, the remaining soldiers of *Dora II* prepared to die in a flourish of nihilistic violence – taking as many of the enemy with them as possible. Throughout 27 April the rapidly dwindling SS company repulsed numerous enemy assaults, but at first light on the 28th the Red Army delivered the *coup de grâce* with two furious infantry assaults. The surviving 11 SS soldiers, Porsche included, having exhausted their ammunition, fixed bayonets and met the second and final Soviet assault with a counter-charge. In

BELOW: *A German anti-aircraft artillery battery guarding defence installations in the Gulf of Finland. The gun is a 2cm Flak 38 light anti-aircraft gun.*

this bitter and bloody clash of hand-to-hand combat the SS Tank Destruction Company *Dora II* was annihilated.

ANTI-TANK MINES

The final category of anti-tank weapons was the anti-tank mine. This had a large explosive charge to disable an enemy tank, generally by damaging a track or a wheel. Studies conducted after the war by the Allies demonstrated that the mine was actually the most cost-effective Nazi anti-tank weapon. The most common anti-tank mine use by the Waffen-SS was the Teller-Mine (TMi). This weapon came in several variants, the earliest being the TMi 29 which appeared in the early 1930s. This 45cm- (18in-) diameter mine carried 6kg (13.2lb) of TNT. By the start of the war, the TMi 35 had replaced it in production, though the TMi 29 continued to see widespread service until the middle of the war. The TMi 35 remained the standard German anti-tank mine until 1943. This smaller 32cm- (13in-) diameter mine was detonated by pressure placed on the lid which fired a central igniter. Troops could also screw additional pull igniters into the base and side and anchor them to the ground to provide anti-lifting capability. In 1943 a cheaper version intended to save on precious raw materials, the TMi 35 Stahl, entered service. Of cruder design, it required greater pressure to trigger. Late in 1943, the TMi 42 entered service. This mine had a smaller pressure plate and a simpler main igniter, though like its predecessors it had a socket for a trip igniter to hamper lifting. It was some 32cm (13in) in diameter and carried Amatol explosive. The last of the Teller-Mines, the TMi 43 Pilz entered Waffen-SS service in 1943, though in much smaller quantities. Easily distinguished by its mushroom-shaped head, it saw only limited service. It was 32cm (13in) in diameter and also carried Amatol explosive. Of even more simplified design, the mine detonated when applied pressure crushed a central sheet metal cap.

On 7 July 1943, during the Battle of Kursk, SS engineers of the *Leibstandarte* employed Teller mines to trap and smash a Soviet armoured regiment engaged in a bitter struggle with a Tiger Tank of the 13th Company, 1st SS Panzer Regiment. As the Tiger's gunner kept the attention of the Soviet tank crews, the SS engineers slipped out from camouflaged foxholes that they had allowed the Red Army tanks to pass by in order to deploy rows of Teller-Mines behind the Soviet tanks. When the T-34 tanks attempted to retire under cover of smoke, they ran over the newly laid Teller-Mines and suffered further heavy losses. Between them, the Tiger and the Teller-Mines accounted for 22 Soviet T-34 tanks.

The Germans also used several more specialised anti-tank mines. The leichte Panzermine, originally developed by the Luftwaffe for easy transportation and use by airborne troops, was packed in specially designed crates of five mines, which could be dropped in batches by parachute without detonating! The 500th and 600th SS Parachute Battalions received limited quantities of this mine which was used, for example, during the raid on Marshall Tito's mountain cave headquarters near Drvar in 1943. The Waffen-SS troops liberally sprinkled the mines to cover their retreat and deter pursuit as they sought to link up with the ground column sent to rescue them.

THE BAR MINE

The Riegelmine 43 (RMi 43) was first encountered by the Allies in Normandy after the D-Day landings. Its unique bar-shaped design meant that many fewer mines had to be laid to form an effective anti-tank barrier, and it soon proved to be the most effective anti-tank mine in German wartime service. Some 80cm (3ft 2in) in length, it weighed 9.6kg (21.1lb) and contained Amatol explosive. Waffen-SS engineers frequently used the mine in tandem or in multiple combinations to deny roads, crossroads and bridges to advancing enemy tanks.

ABOVE: *A 2cm Flak 38 light anti-aircraft gun in use with SS police units thrown into battle on the Eastern Front during February 1945.*

Waffen-SS troops also employed various anti-tank grenades. One of the more common was the Panzerwurfmine 1 (Luft), or PzWuMi 1(l). This was a hand-held mine that was thrown by means of a wooden handle up into the air, like a stick grenade, and which utilised a hollow-charge grenade fitted with four folding vanes, almost like a mini-parachute, that helped guide it towards its target. The vanes gave the mine an effective range of 35–40m (38–44yds), but accuracy decreased with distance. It could penetrate 64mm- (2.5in-) thick armour and the actual warhead resembled that of the stick grenade. Unlike the latter, however, the PzWuMi 1(l) had to be thrown under-arm while holding the ends of the vanes to ensure that they deployed properly in flight.

Waffen-SS engineers also employed a variety of demolition charges against tanks. Some of these so-called 'satchel' charges were magnetised to allow them to stick to the sides of tanks. The larger charges varied from between 8–50kg (19.6–110lb) and were typically simple rectangular metal boxes filled with appropriate explosive.

Then there was the Haft-Hohlladung, a 3kg (6.6lb) magnetic anti-tank assault weapon filled with a hollow-charge warhead which was placed on the side of enemy tanks and then ignited by a short-delay detonator. SS-Unterscharführer (Sergeant) Emil Dürr, a squad leader in the 4th Company, 26th SS Panzergrenadier Regiment, 12th SS Panzer Division *Hitlerjugend*, was posthumously decorated with the Knight's Cross for outstanding bravery when using a Haft-Hohlladung against Canadian armour near Caen on 26 June 1944. Dürr leapt from cover, being wounded in the process, to attach the magnetic charge to the side of a passing tank that was about to overrun his unit's position. But the magnetic charge would not adhere; three times Dürr attached the charge and thrice it fell off. Instead, Emil Dürr pulled the detonator and held the charge against the side of the tank, blowing both himself and the tank to pieces.

At Lushno on 26 September 1941, the *Totenkopf* Division employed special 12-man tank-destruction units equipped with satchel charges, mines, grenades and Molotov cocktails against masses of Soviet armour. One such group was led by SS-Hauptsturmbannführer (Captain) Max Seela. To set an example to his men and demonstrate the proper techniques of anti-tank combat, Seela destroyed the first Russian tank that had broken through the German lines by placing a double satchel charge against the turret and detonating it by lobbing a grenade onto the tank. That day Seela's squad knocked out seven enemy tanks and then mercilessly gunned down their crews as they sought to escape from their doomed vehicles, including even those attempting to surrender. The brutality displayed by the SS troops that day epitomised the savage, bitter struggle for survival that characterised warfare on the Eastern Front.

The appearance of the tank on the modern battlefield necessitated providing the infantry with some effective counter to armour. But since anti-tank guns could only be produced in limited quantities, German infantry might often have to face enemy armour without anti-tank support. To guard against this eventuality, the Germans provided the grenadier with the rifle grenade, a simple but effective weapon, at least in the early stages of the war. This was a hollow-charge anti-tank grenade designed to be fired from a standard 7.92mm German rifle. In combat, the rifleman fitted a special adaptor to the rifle and inserted a blank cartridge into the breech to provide the propellant charge. These early grenades had limited range, about 50m (55yds), and could only penetrate 30mm (1.2in) of armour. This was sufficient in the earliest stages of the war, but even by 1940 most tanks were all but invulnerable to the rifle grenade, unless it managed to hit and damage

BELOW: *A camouflaged 2cm Flakvierling 38 quadruple-barrelled self-propelled anti-aircraft gun mounted on the back of a halftrack artillery tractor.*

Flakvierling 38

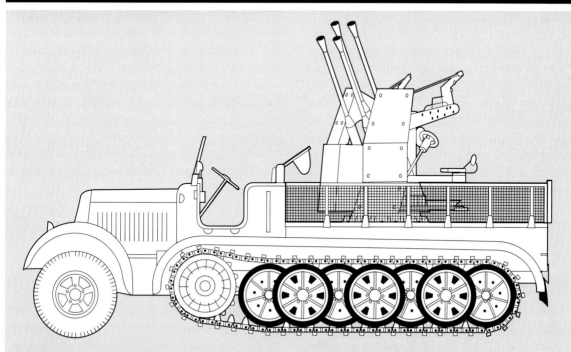

German Designation: 2cm
  Flugabwehrkanonenvierling (Flakvierling) 38
Weapon Type: Four-Barrelled Light Anti-aircraft
  Gun
Calibre: 2.cm (0.8in)
Length (of piece): 2.25m (7ft 5in)

Weight: 1509kg (3327lb)
Traverse: 360 degrees
Elevation: -10 to +100 degrees
Muzzle Velocity: 900m/s (2951ft/s)
Maximum Ceiling: 2200m (2407yds)
Rate of Fire: 4 x 200–450rpm = total 800–1800rpm

vulnerable parts like the engine cover or tracks. Later, the Germans introduced heavier grenades with longer range that could penetrate 90–126mm (3.5–5in) of armour. As the war progressed, though, and enemy tanks became better protected, the rifle grenade became increasingly ineffective as an anti-tank weapon, and it was largely relegated to use against soft-skinned vehicles, such as trucks and jeeps, which it could easily neutralise.

SS-Sturmbannführer (Major) Joachim 'Jochen' Peiper of the *Leibstandarte* Division and notorious for the Malmédy Massacre,

became renowned for his personal courage and fearlessness in the brutal struggle for survival that materialised on the Eastern Front. It is said that after knocking out a Soviet T-34 tank with a rifle grenade at point-blank range, Peiper turned to his comrades and quipped, 'Lads, that should do for the close-combat badge', a medal awarded for hand-to-hand infantry fighting rather than anti-tank combat! A Waffen-SS anti-tank weapon of last resort was the direct fire of artillery over open sights, which often proved effective in an emergency anti-tank role, a powerful if dangerous measure.

Anti-aircraft guns are designed to deliver a barrage of exploding shells – flak – against enemy aircraft. The effectiveness of such flak fire lays less in its ability to bring down enemy aircraft – to do so generally required firing thousands of rounds – than in distracting enemy planes so that they could not complete their attack successfully. The first SS flak unit emerged in 1939 when the SS raised an anti-aircraft machine-gun battalion. After the Polish Campaign, the newly activated *SS-Verfügungs* Division received a three-company light anti-aircraft battalion equipped with some 36 2cm self-propelled anti-aircraft guns mounted on Sdkfz 10 halftrack artillery tractors. In July 1940, the *Leibstandarte* received a self-propelled light flak battery which expanded to battalion size during the spring of 1941. Both the *Totenkopf* and the new SS Motorised Infantry Division *Wiking* each received similar self-propelled anti-aircraft battalions. By the start of Operation 'Barbarossa' on 22 June 1941, the number of SS anti-aircraft batteries had risen to 16.

EQUIPPING THE SS FLAK UNITS

From then on the number of SS flak units grew steadily. In July 1941, the Waffen-SS received its first medium battery equipped with 3.7cm guns, and during 1942–43 SS panzergrenadier regiments received regimental light flak companies. During the middle of the war new weapons joined the SS flak arsenal, including self-propelled 37cm medium anti-aircraft guns and quadruple 2cm guns. By 1944, SS mechanised formations had become well-equipped with flak guns. Besides a five-battery motorised flak battalion, divisions also had additional anti-aircraft platoons and companies in their panzergrenadier, panzer and artillery regiments. The 1944 SS panzer division was authorised 80 towed and 40 self-propelled 2cm guns, six 2cm quad weapons, nine 3.7cm guns and 12 heavy 8.8cm flak pieces. While less well-equipped, SS

grenadier and mountain divisions still possessed 18 towed and 12 self-propelled 2cm guns. In total during the war, the Waffen-SS raised 49 anti-aircraft battalions and 22 independent companies.

The first anti-aircraft gun to see service in the Waffen-SS was the 2cm Flak 30 which saw action in Poland in 1939. It was quick-firing (280 rounds per minute), weighed only 483kg (1065lb) and came on a three-legged platform that was transported on a two-wheel trailer. It fired 0.12kg (0.26lb) high-explosive, incendiary or armour-piercing rounds at a muzzle velocity of 899m/s

(2950ft/s) to a maximum vertical range of 2134m (2345yds) and to 2697m (2950yds) when used in a ground role. After the Polish Campaign an improved gun, the 2cm Flak 38 entered service. Among its enhancements were greater range and an increase in rate of fire to an impressive 450 rounds per minute. This gun became the primary light anti-aircraft gun used by the Waffen-SS for the rest of the war.

The 2cm Gebirgsflak 38 was a modified version of the basic 2cm Flak 38 intended for use by mountain units in alpine terrain. As such it was much lighter and more portable than the standard version. It served with SS mountain formations like the 21st SS Volunteer Mountain Division *Skanderbeg* (Albanian No. 1), as well as in a number of SS grenadier divisions in place of the standard Flak 38. The authorised allotment for this anti-aircraft weapon was once again 12 guns per company.

BELOW: *This 3.7cm Flak 36 anti-aircraft gun, mounted on the back of an artillery tractor, is supporting ground forces during the fighting for Vitebsk in 1944.*

Flak 18

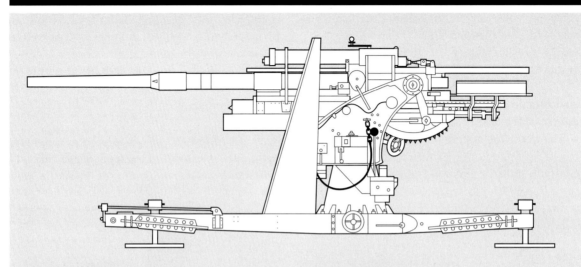

German Designation: 8.8cm Flugabwehrkanone (Flak) 18

Weapon Type: Heavy Anti-aircraft Gun

Calibre: 8.8cm (3.5in)

Length (of piece): 4.93m (16ft 2in)

Weight: 3710kg (8180lb) = 3.7 tonnes (3.65 tons)

Traverse: 360 degrees

Elevation: -3 to +85 degrees

Muzzle Velocity: 820-840m/s (2733-2754ft/s)

Maximum Ceiling: 10,600m (11,596yds)

Rate of Fire: 15–20rpm

A formidable, though thankfully for the Allies uncommon, light anti-aircraft weapon was the Mauser-produced 2cm Flakvierling 38, a quadruple mounting of the Flak 38. Developed in 1940 for the German Navy, it was quickly adopted by the SS. It weighed 1520kg (3352lb) and its four barrels could pour forth a lethal 1800 rounds per minute, making it deadly, not just against low-flying aircraft but also against ground targets. It was issued in platoon strength to the headquarters companies of SS armoured artillery and panzer regiments. It also served occasionally at corps level. The 101st SS Flak Company of the premier I SS Panzer Corps, for example, contained two platoons each equipped with three quad 2cm self-propelled guns.

The 3.7cm Flak 18 introduced in 1935 was the first medium gun to join the SS inventory. It weighed 1757kg (3858lb), was mounted on a cruciform platform and fired a 0.56kg (1.2lb) shell to a vertical height of 4785m (5235yds) or to 6492m (7100yds) against ground targets. Its rate of fire was 160 rounds per minute. However, it proved heavy and cumbersome, and an improved 3.7cm Flak 36 soon superseded it. Modifications reduced the weight of the Flak 36 to 1544kg (3504lb), but performance remained the same as its predecessor. In 1943 a further refined 3.7cm Flak 43 appeared. Combining reduced weight – 1247kg (2750lb) – with an increased rate of fire of 250 rounds per minute, it remained in production until the end of the war. The twin-mounted 3.7cm Flakzwilling 43 was also built, with the barrels mounted vertically above one another like a shotgun. It weighed 2781kg (6132lb) and had a rate of fire of 500 rounds per minute. Very few are believed to have seen service in the Waffen-SS, however.

The most famous German anti-aircraft gun of World War II was the heavy 'Flak 88'. The earliest model, the 8.8cm Flak 18 L/56, entered service in 1934 and was the result of inter-war collaboration between Krupp and Bofors of Sweden. It was of novel design, with the gun mounted on a cruci-form platform from which it fired with outriggers extended, and the barrel mounted on a tall central pedestal which permitted elevation up to 85 degrees. The gun weighed 4985kg (10,992lb) and fired a 9.4kg (20.7lb) shell at a muzzle velocity of 795m/s (2609ft/s) firing armour-piercing rounds against ground targets and 820m/s (2691 ft/s) fir-ing high-explosive rounds against aircraft. It had a rate of fire of 15 rounds per minute and a maxi-mum vertical range of 8000m (8758yds). Deployed to Spain in 1936 as part of the German Condor Legion, its high velocity and flat trajectory made the gun very accurate and effective in both an anti-tank and artillery role.

In the late 1930s, the slightly modified 8.8cm Flak 36 went into production. The only signifi-cant difference being the addition of replaceable barrel liners that reduced wear, as well as a three-part barrel assembly. It weighed 4985kg (10,992lb) without limbers. Deployed in a ground role it had a maximum range of 14,680m (16,060yds), although it was rarely used above 2000m (2187yds) against enemy armour. At this range it could penetrate 8.8cm (3.5in) and 7.2cm (2.8in) of vertical and 30-degree sloped armour respectively.

In 1939, the 8.8cm Flak 39 entered service with a two-part barrel and gun-laying enhance-ments. In 1941 a much more refined gun, the 8.8cm Flak 41, went into production. This gun was specifically designed for a dual role and thus possessed a genuine anti-tank capability. Its

BELOW: *A 3.7cm medium anti-aircraft gun of an SS flak battery in action in Russia. Note the range finder being used by the gunner on the left.*

LEFT: *German engineers rig a series of Teller-Mines, including what appear to be TMi 35 and 35 Stahl, to destroy a building during a retreat west in Russia.*

longer 71-calibre barrel, as compared to 56 calibres on the Flak 36, gave it an increased muzzle velocity of 980m/s (3216ft/s) firing armour-piercing rounds and a horizontal range of 20,000m (21,880yds)! The gun's penetration was 132mm (5.2in) at 2000m (2189yds). At the same time, streamlining reduced its overall height to 2.36m (7ft 9in), the high silhouette of former models being their most significant drawback in a ground role. In service it proved a very versatile weapon, and it continued in production until 1945. Although heavier than it predecessors – it weighed 7800kg (17,199lb) – this was more than compensated for by the enhanced performance, lower silhouette and superior rate of fire of 20 rounds per minute.

THE DEMAND FOR FLAK 88S

Early in 1941 the SS received its first heavy flak unit, when the 6th Battery, SS Artillery Regiment *Leibstandarte*, received three 8.8cm Flak 18 heavy anti-aircraft guns. During the German conquest of Greece, one of these guns established a new tank-destruction distance record when it knocked out a British tank at a range in excess of 6000m (6568yds)! This achievement demonstrated unequivocally the weapon's outstanding anti-tank capabilities. Thereafter, SS troops clamoured for more Flak 88s. In May 1941 a second heavy battery reached the *Das Reich* Division, followed by a third during August 1941.

During 1942 the premier SS divisions raised four- or five-battery anti-aircraft battalions each of which contained two or three heavy batteries. That same year new battalions joined the *Totenkopf*, *SS-Polizei* and *Florian Geyer* Divisions. During 1943 the *Prinz Euge*n, *Hohenstaufen*, *Frundsberg* and *Nordland* Divisions received similar battalions, as did the

Hitlerjugend, Reichsführer-SS, Götz von Berlichingen and *Horst Wessel* Divisions in 1944. Only one new SS flak battalion, however, appeared in the spring of 1945, attached to the 32nd SS Grenadier Division *30 Januar.*

It was élite SS anti-aircraft gunners who guarded Hitler's summer residence and headquarters at Obersalzburg on the Bavarian-Austrian frontier. Part of the SS 'Special Forces', they endeavoured to keep Hitler's dwelling safe from the ravages of the Allied air forces. In April 1945, the unit reorganised for ground combat as a mountain battalion, and was one of the core units around which Hitler hoped to construct the mythical Alpine redoubt entitled the 'Fortress Alps'.

One of the most decorated wartime SS flak gunners was SS-Unterscharführer (Sergeant) Heinrich Gottke, a forward observer with the 17th SS Flak Battalion *Götz von Berlichingen.* Stationed at Erfweiler near the Reich's western border on 16 December 1944, he observed an American armoured force breaking through the division's frontline and advance on his observation post. With no means to oppose the enemy tanks, Gottke calmly called down the fire of his battalion on his own position with the phrase 'Fire on the hermit'! The barrage halted the enemy advance and for his bravery and disregard for his own personal safety Gottke was awarded the Knight's Cross, one of only three flak gunners ever to receive the award.

The last SS flak unit to see combat during the war was the SS Kampfgruppe *Dirnagel*, which was formed on 25 March 1945 at Munich from the 1st SS Flak Training and Replacement Regiment. Led by SS-Obersturmbannführer (Lieutenant-Colonel) Dirnagel, the battle group comprised one battalion of gunners remustered as infantry and a five-battery flak battalion. The regimental group fought briefly in front of Munich in late March and early April, before joining the German retreat into Austria to make a determined last stand in the 'Fortress Alps'. Hitler's suicide in Berlin, however, released the SS troops from their obligation to make a suicidal last stand, and the remnants of the regiment capitulated in Austria in early May 1945.

Teller-Mine 35

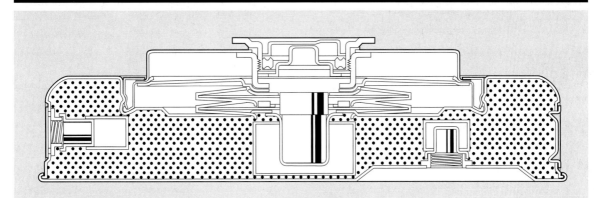

German Designation: Teller-Mine (TMi) 35
Weapon Type: Anti-tank Mine
Explosive: TNT
Firing Pressure: 80-180kg (177-397lb)

Height: 8cm (3in)
Diameter: 32cm (1ft 1in)
Weight: 8.7kg (19.2lb)

CHAPTER 4

Armoured Cars and Halftracks

Armoured cars and halftracked armoured personnel carriers were often harbingers of the arrival of the Waffen-SS on the battlefield. Not only did they provide the mobility that was crucial to success, they became synonymous with the daring ethos of the Waffen-SS.

As pioneers of mechanisation in the inter-war period, the German Army used armoured cars widely. One reason for this was that the 1919 Treaty of Versailles which ended World War I had grudgingly allowed the German military to retain small numbers of armoured cars for internal security and police duties. The Germans were thus able to gain early practical experience in the development and use of this type of equipment. As a result, after Hitler's rise to power Germany quickly introduced three series of modern armoured cars intended for reconnaissance and screening roles within the new panzer divisions that emerged in the late 1930s.

LEFT: *An Sdkfz 251 armoured personnel carrier with a 7.92mm MG 34 machine gun mounted in a full forward shield. Note additional kit lashed to the sides.*

The first German armoured cars developed were the heavy six-wheeled Sdkfz 231 series, 928 of which were produced during 1930–35. This was followed by the light four-wheeled Horch Sdkfz 221 series, 2118 of which were manufactured between 1935 and 1942. The third series were the Bussing-NAG eight-wheelers, confusingly also designated Sdkfz 231, some 1235 of which were built during 1937–42. A second generation of armoured cars, the Sdkfz 234 series, entered service in 1943 and over 1000 were completed by the end of the war. Despite both the quality of these German weapons and the popular image of fast-moving armoured cars as being integral to the early German Blitzkrieg victories, production was limited and these vehicles remained rare throughout World War II. On 1 November 1944, for example, there were only

562 armoured cars in the entire German armed forces!

A brief examination of the armoured cars in Waffen-SS service at the start of the war illuminates this reality. In September 1939, the four SS motorised infantry regiments each possessed an armoured car platoon that deployed eight Sdkfz 221 and 222 light armoured cars. These vehicles belonged to the series of light four-wheeled

BELOW: *An Sdkfz 221 light armoured car of the* Das Reich *Division in the Balkans in 1941. The man in the raincoat is probably a local collaborator.*

armoured cars manufactured in the 1930s – the Sdkfz 221, 222 and 223 – that differed from one another in weight, armament and turret construction. The four-wheeled Sdkfz 221 went into production in 1935. This 3.75-tonne (3.7-ton) compact armoured car was powered by a 75bhp Horch petrol engine that delivered an impressive road speed of up to 80kph (50mph). A 100-litre (22-gallon) fuel tank also provided a good operational radius of 280km (174 miles) by road and 200km (124.5 miles) cross-country. The vehicle was operated by a two-man crew and mounted a single 7.92mm MG 34 machine gun as its armament. Armour ranged from 14.5mm (0.6in) on the front to 8mm (0.3in) on the sides and rear.

THE SDKFZ 222

The four-wheeled Sdkfz 222 was a heavier 4.8-tonne (4.7-ton) version of the Sdkfz 221 that carried a 2cm KwK 30 cannon and co-axial 7.92mm MG 34 mounted in a 10-sided, open-topped rotating turret. A wire-mesh hood protected the turret top against grenade attacks. Some late-production models carried a taper-bore 2.8cm heavy anti-tank rifle. A rarer variant was the Sdkfz 223 radio car, a 4.4-tonne (4.3-ton) four-wheeled vehicle armed with a 7.92mm MG 34 that carried extra communications equipment and an additional third crew-member who operated the radios. The Sdkfz 223 was easily distinguished from the Sdkfz 221 by the rectangular frame aerial mounted on top of the vehicle. This aerial could be folded back and down to give the machine gunner an improved field of fire. All three of these four-wheelers possessed a wheelbase that matched the standard German rail gauge. They could thus run on any standard railway track with their tires removed, and could be driven from either of the driving positions fitted at each end of the vehicle.

In September 1939, the SS-Verfügungstruppe also deployed an independent motorised reconnaissance battalion which fought in Poland,

Sdkfz 221

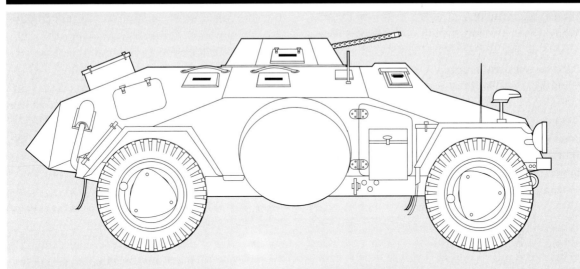

German Designation: Sdkfz 221 leichter
  Panzerspahwagen (lePSW)
Vehicle Type: Light four-wheeled Armoured Car
Crew: two
Weight: 3.75 tonnes (3.8 tons)
Length: 4.8m (15ft 9in)
Width: 1.945m (6ft 5in)
Height: 2.15m (7ft 1in)
Armament: 1 x 7.92mm (0.312in) MG 34
Ammunition Stowage: 2000 rounds
Armour:
  -Front (Nose): 14.5mm (0.6in)

-Front (Driver's Plate): 14.5mm (0.6in)
-Sides: 8mm (0.3in)
-Rear: 8mm (0.3in)
Engine: 75 bhp Auto Union Horch V-8 cylinder
  petrol
Fuel capacity: 100 litres (22 gallons)
Maximum Road Speed: 80kph (50mph)
Maximum Cross-Country Speed: 32 kph (20mph)
Operational Range (Road): 280km (174 miles)
Operational Range (Cross-Country): 200km
  (124.5 miles)

attached to the ad hoc Panzer Division *Kempf* formed in East Prussia from the mechanised units shipped their to participate in the planned autumn army manoeuvres. The battalion possessed 11 armoured cars: an armoured signals platoon equipped with a single example each of the four-wheeled Sdkfz 260 and 261 armoured radio cars, a solitary six-wheeler Sdkfz 263 signals vehicle, and an armoured car platoon with eight Sdkfz 221 and 222 light armoured cars. The SS motorised anti-tank battalion also fielded a solitary Sdkfz 261 armoured radio car. In total, at

the outbreak of hostilities, therefore, the front-line militarised SS therefore fielded a mere 44 armoured cars!

The four-wheeled Sdkfz 260 and 261 armoured radio cars were variants of the basic Sdkfz 221 that carried different configurations of radios and antennae for communications work. They were open-topped and carried no armament. The six-wheeled, turretless Sdkfz 263 radio car was a variant of the first series of German armoured cars constructed: the 5-tonne (4.9-ton) Mercedes Sdkfz 231 six-wheeler heavy armoured car which went into production in 1930. A 65bhp six-cylinder Daimler-Benz M 09 petrol engine powered the Sdkfz 231 up to 1932; subsequent cars featured a Bussing 60bhp water-cooled petrol engine. The final Sdkfz 231 production variant, manufactured from 1934 by Magirus, was heavier at 5.3 tonnes (5.3 tons) and featured the company's own S-88 six-cylinder 70bhp petrol engine. All three Sdkfz 231 variants possessed similar performance: a maximum road speed of 65kph (40mph), together with an effective range of 300km (186 miles) by road and 140km (87 miles) cross-country. The armoured car carried a crew of four, mounted a 2cm KwK 30 cannon in a fully traversing turret, and possessed 14.5mm (0.6in) frontal armour with 8mm (0.3in) plates on the sides and rear.

The Sdkfz 232 was another variant of the Sdkfz 231 six-wheeled heavy armoured car that carried extra radios and a distinctive large overhead frame aerial for signals work. The Sdkfz 233 was Germany's first wheeled heavy support vehicle. This armoured car mounted a limited traverse 7.5cm StuK L/24 short-barrelled gun in an open fighting compartment, which replaced the original turret of the Sdkfz 231.

BATTLEFIELD SURVIVEABILITY

The third series was the confusingly designated eight-wheeled Bussing-NAG Sdkfz 231 heavy armoured cars. Though similar in design to the Sdkfz 231 six-wheelers, the eight-wheeled Sdkfz 231 armoured cars were longer and carried the same turret mounted farther forward. German firms also constructed small numbers of eight-wheeled Sdkfz 232 and 263 signals cars and Sdkfz 233 heavy support vehicles as well. These eight-wheeled radio cars were easily distinguished from their smaller cousins by their more forward-mounted frame aerials. What made the eight-wheelers unique among German armoured cars, though, was that all eight wheels were individually powered. This ensured that the vehicle

could still move even after enemy fire had damaged one or more wheels! These eight-wheelers became the most common German heavy armoured car of the war.

Heavy armoured cars first entered SS service during the winter of 1939–40. By May 1940 the *SS-Verfügungs* Division had received six Sdkfz 231 and 232 heavy armoured cars. Yet by this date, the number of armoured cars in Waffen-SS service had only increased to 66 vehicles. The SS *SS-Verfügungs* Division deployed 55 of these in its three regimental platoons and motorised reconnaissance battalion, while the *Leibstandarte* fielded a platoon of eight light armoured cars. The *Totenkopf* Division was supposed to follow the same organisation as the *SS-Verfügungs* Division, but production delays prevented this and it fielded just three armoured radio cars. Even by June 1941 the *Totenkopf* still possessed only a single platoon of eight Sdkfz 221 and 222 four-wheeled light armoured cars.

A SLOW EXPANSION

It was only after the onset of Operation 'Barbarossa', the German invasion of Russia in June 1941, that the Waffen-SS began to receive greater numbers of armoured cars. During 1942, the *Totenkopf* received three armoured car platoons and the *Leibstandarte* a full company. But it was not until 1943 that these vehicles became more bountiful. That year the *SS-Polizei*, *Wiking*, *Florian Geyer* and *Nordland* Divisions each received full-strength armoured car companies. That same year V SS Mountain Corps was also was allocated a heavy platoon of six armoured cars, and the 13th SS Mountain Division *Handschar* gained a light platoon of six vehicles.

The Waffen-SS envisaged armoured cars conducting reconnaissance to obtain battlefield intelligence. Unlike their Anglo-American counterparts, for example, who utilised a range of intelligence sources, the Germans relied heavily on battlefield intelligence and were always willing to fight for information. In general, therefore, Waffen-SS armoured reconnaissance battalions were both well-equipped with armoured cars and halftracks. It also became the credo of Waffen-SS armoured reconnaissance units that audacity and speed, as well as superior training, tactics and morale, allied to German aerial superiority over the area, could offset limitations in fire-

power and protection to allow SS armoured cars to acquire valuable intelligence on the battlefield.

Armoured reconnaissance battalions, therefore, invariably spearheaded Waffen-SS offensives with aggressive verve. In the East during

BELOW: *An Sdkfz 222 light armoured car of the 8th SS Cavalry Division* Florian Geyer *(identified by the division's sword insignia on the front of the vehicle).*

ABOVE: *An SS eight-wheeled Sdkfz 231 heavy armoured car in France, May 1940. This vehicle was armed with a 20mm gun and 7.92mm machine gun.*

September 1943, for example, the 1st Armoured Car Company, SS Armoured Reconnaissance Battalion *Das Reich*, reconnoitred a suspected Soviet armoured build-up in a village on the exposed flank of the division. The company's heavy platoon, led by SS-Untersturmführer (2nd Lieutenant) Alfred Siegling, was equipped with six Sdkfz 231 eight-wheelers. This platoon resolutely entered the village – despite the known presence of Soviet armour – to ascertain the enemy's strength and intentions. To the astonishment of the surprised Russians, Siegling's patrol roared through the village at top speed, passing within metres of dozens of Soviet T-34 tanks

parked off the village square. But the Red Army troops were so surprised by Siegling's daring sortie that they failed to react quickly enough. By the time the Soviet tank crews had rejoined their tanks, started their engines and trained their guns, all they could see was the dust thrown up by the departing German armoured cars as they exited the far side of the village unscathed! Siegling provided an exact count of the Soviet tank strength, and the division was able to adjust its dispositions to meet and repulse the subsequent Soviet attack. For his daring and successful feat of arms, Alfred Siegling received the Knight's Cross.

Siegling's example illustrated the increasing reliance SS armoured reconnaissance units had to place during 1942–43 on stealth, speed and surprise as ever-more powerful enemy weapons entered service on the Eastern Front. Indeed, by 1943 the Germans finally recognised the obsolescence of their first-generation, 1930s-vintage armoured cars. The marked increase in the power of tank and anti-tank guns rendered vulnerable these thinly protected German armoured cars, which also lacked the firepower to fight their way out of danger. Consequently, during 1943–44 the Waffen-SS progressively withdrew their surviving light armoured cars to second-line units, where they provided valuable service in the bitter anti-partisan war the Germans faced in much of occupied Europe, especially throughout Eastern Europe and the Balkans.

ENTER THE PUMA – DEADLY CAT

The Germans realised that they needed a second generation of more powerfully armed and protected armoured cars that could survive on the modern battlefield. The result was the Sdkfz 234 series of heavy eight-wheeled armoured cars. Developed during 1942, the Sdkfz 234 was purpose built to endure the harsh, rugged conditions on the Eastern Front that had disabled so many of the earlier four-wheelers that lacked independent drive. German firms produced the Sdkfz 234 armoured car in four variants. The basic Sdkfz 234/1 weighed 11.7 tonnes (11.65 tons) – much heavier than any previous German armoured car – and its powerful 210bhp, 12-cylinder Tatra 103 air-cooled diesel engine generated an impressive top speed of 90kph (56mph). The 234/1 carried a 2cm cannon as its main armament, by 1942 the minimal armament necessary for survival on the battlefield. During 1943 this model was issued to the armoured car companies of SS panzer and panzergrenadier divisions. In combat, it proved robust and swift enough to avoid more powerfully armed opponents.

However, by 1943 the 2cm gun of the Sdkfz 234/1 was simply not powerful enough to enable the vehicle to fight in order to obtain battlefield intelligence. This need for greater firepower led to the development of the Sdkfz 234/2 Puma, one of the most elegant and best-known armoured cars of World War II. In reality, however, the Puma remained a rare vehicle. Widely regarded as the best armoured car of World War II, it mounted a long-barrelled 5cm KwK L/60 gun in a fully revolving turret. The heavier gun and turret slightly decreased the Puma's speed and range, but these limitations were more than compensated for by the vehicle's proven anti-tank ability. From late 1943 Pumas began to join SS armoured reconnaissance battalions, where they quickly acquitted themselves well. Himmler ultimately hoped to replace the Sdkfz 234/1 completely with the Puma, but the increasing diversion of resources to other assembly lines ensured that this was impossible, and both Sdkfz 234 vehicles continued to serve alongside each other until the end of the war.

MORE AND MORE VARIANTS

The success of the Puma led Hitler to order the production of the Sdkfz 234/3, a more heavily armed battlefield fire-support version of the Puma. Essentially an updated version of the earlier Sdkfz 233, it carried the 7.5cm KwK L/24 low-velocity gun formerly mounted in pre-1941 Panzer IV tanks. This vehicle was open-topped and the gun's traverse was limited. The vehicle weighed 9.95 tonnes (9.8 tons) combat-laden, including 55 rounds of ammunition. The marrying of this gun to the Sdkfz 234 armoured car was obviously expedient, since the ill-protected crew preferred not to expose them themselves to the dangers of close combat, but instead sought to provide fire support from a distance, hopefully out of range of enemy weapons.

Even more illustrative of the expedient nature of late-war German equipment was the final vehicle in the series, the Sdkfz 234/4, which emerged in 1944. It featured the more potent 7.5cm KwK L/48 gun mounted in an open-topped superstructure. The gun was simply placed straight into the open fighting compartment and bolted in place. The result was a crude, albeit potent, eight-wheeled, limited-traverse tank destroyer. Even though the vehicle was only built in small numbers, it still provided potent anti-tank firepower at a time when the Waffen-SS desperately needed mobile anti-tank weapons to counter the ever-growing quantities of enemy armour on the battlefield.

LIGHTNING DASH AT ARNHEM

The emergence of the Sdkfz 234 once again provided Waffen-SS armoured reconnaissance units with the tools to resume their preferred doctrinal role. But even with the Sdkfz 234, the 1944 battlefield remained a dangerous place for armoured cars. For although the Waffen-SS continued to use its armoured cars with aggressive élan, by this phase of the war such offensive spirit rarely proved sufficient to offset the growing battlefield superiority of the Allies. Illustrative of this grim reality was the combat experience of the 9th SS Armoured Reconnaissance Battalion *Hohenstaufen* during the Battle of Arnhem. On the morning of 17 September 1944, the very day of the Allied airborne landings, the 30-year old battalion commander, SS-Hauptsturmführer (Captain) Viktor Gräbner, was presented with the Knight's Cross in a ceremony at Hoenderloo 16km (10 miles) north of Arnhem. For this occasion, Gräbner's entire battalion was lined up in parade formation to review past their commander – until a vast Allied airborne armada flew overhead. As parachutes filled the skies above him, the commander of II SS Panzer Corps, SS-Obergruppenführer (General) Wilhelm Bittrich – who had just pinned the Knight's Cross medal on Gräbner's chest – ordered his subordinate to reconnoitre south towards Nijmegen with his force of 40 armoured cars and armoured personnel carriers (APCs).

At 1800 hours on 17 September, Gräbner's battalion thundered unchallenged across the Arnhem road bridge heading south. Soon after the battalion had crossed the bridge the following unit – the 1st Company, 10th SS Armoured Reconnaissance Battalion *Frundsberg* – ran into the middle of a surprise assault by the 2nd British Parachute Battalion that seized control of the bridge. News of the fall of the bridge reached Gräbner outside Nijmegen, and he immediately headed back to Arnhem that evening. The following morning, Gräbner, a man of action, attempted to retake the Arnhem bridge from the south using surprise and deception. At 0900 hours he personally led – in a British Humber armoured car he had captured in Normandy – his battalion's 22 armoured cars and halftracks in column formation on a 7000m (7663yds) dash for the bridge. He hoped that the familiar silhouette of the Humber might fool the defenders into believing that the column approaching was the British ground force intended to relieve them. But behind the Humber trundled the battalion's Sdkfz 234/2 Pumas that were to provide the firepower necessary to take the bridge. Once the defenders saw through the German ruse, as they eventually must, the game would be up.

GRÄBNER'S GAMBLE

At top speed, Gräbner's Humber roared forward towards the bridge. At first his *ruse de guerre* worked, and he was very close to the bridge before the British realised what was happening and opened up on the German column with everything they had. Gräbner got five vehicles onto the bridge before the sixth hit a mine, blew up and completely blocked the bridge. With his retreat cut off, Gräbner's attack withered and died in the face of determined resistance by the dug-in British paras, who fired PIAT anti-tank rounds, hand grenades and mortar rounds into

RIGHT: *An Sdkfz 231 heavy armoured car of the 1st SS Panzer Division* Leibstandarte *amid the ruins of Caen during the fighting in Normandy in June 1944.*

the trapped SS column. By midday Gräbner's attack had faltered completely, his battalion having lost 12 of its 22 armoured vehicles, including six of his precious Pumas. His gamble also proved a costly personal failure, as Gräbner himself was killed at the head of his battalion on the bridge.

Such was the enormous Waffen-SS demand for armoured cars early in the war that it proved impossible for Himmler to acquire sufficient German vehicles through normal procurement channels. The Waffen-SS therefore 'appropriated' many French armoured cars captured in 1940, particularly the versatile Panhard 178, designated in German service as the schwere Panzerspahwagen P204(f). This was an excellent general-purpose armoured car. It weighed 8.3 tonnes (8.2 tons) and carried a 105bhp Panhard ISK petrol engine that gave a top road speed of 72.5kph (45mph). The vehicle mounted a French 2.5cm anti-tank gun as its main armament, a gun that had superior performance to the German

2cm KwK 38. In total, the Waffen-SS acquired 190 Panhards and issued them to their armoured reconnaissance units during 1940–42. Most were, however, destroyed in the bitter fighting on the Eastern Front, and by 1944 the relatively few that remained in SS service had long been relegated to rear-area security tasks.

A vehicle developed in the late 1930s ultimately took on many of the functions originally intended for the armoured car. This vehicle was the halftrack APC, which has become the univer-

LEFT: *An Sdkfz 232 armoured radio car of the* Totenkopf *Division near Arras, May 1940. The car on the right bears the division's symbol.*

sal transporter of infantry units in modern armies. The idea of an armoured troop carrier originated from the British during World War I, who sought to carry infantry forward immune from machine-gun fire. From this concept the tank evolved during 1915–16, and by 1918 the British had converted a number of tanks to unarmed infantry carriers. Peacetime brought an end to such innovation, though the inter-war British Experimental Mechanised Force revealed the potential for carrier-borne infantry to assist armour on the modern battlefield.

The origins of the German halftrack APC thus lay elsewhere – with a French engineer called Adolphe Kegresse – who managed the Russian Tsar's personal motor fleet in the early 1900s. Kegresse made a bogie assembly with rubber tracks and used it to replace the rear wheels on one of the Tsar's vehicles for improved winter traction in snow and ice, thus giving birth to the halftrack. In 1917 Kegresse returned to his native France, and Citroën took up his design to produce a series of Citroën-Kegresse halftracks that served in the inter-war French, British, and American armed forces.

EARLY GERMAN HALFTRACKS

The Germans quickly copied these ideas and introduced very small numbers of Daimler Marienwagen halftracks in 1918. This vehicle was built on the chassis of a Daimler truck, with an Erhardt armoured car body mounted on top and a simple rubber band style of track unit. But Germany completed only four Marienwagens before the end of World War I. The 1919 Treaty of Versailles inadvertently provided further impetus to German halftrack design since it allowed the German police to retain 105 armoured troop carriers. These were deemed necessary to help maintain domestic law and order in the violent times that occasioned Germany's turbulent transition to a democratic republic.

In 1926 the German War Ministry sought to procure a halftracked vehicle for future infantry use. It issued contracts for six halftracks of various weights which led to the emergence of the series of German halftrack artillery tractors that served throughout World War II. The two lightest tractors evolved further into the two standard German halftracked APCs of World War II: the Sdkfz 250 and 251.

THE UBIQUITOUS SDKFZ 251

In 1935, with the first tanks in production and the panzer divisions already in formation, the Germans deemed the 3-tonne (2.95-ton) halftrack tractor chassis suitable in size to carry the then standard 10-man German rifle squad. Preliminary development revealed that only minimal changes to the basic tractor design were needed, as well as the addition of a specially developed faceted, well-sloped armoured superstructure similar to those fitted on German armoured cars. The vehicle possessed rear doors and the crew compartment was left open so that troops could disembark over the sides. The superstructure was protected by 14.5mm (0.6in) of armour plate on the front and 8mm (0.3in) on the sides and rear. The prototype was completed in 1938 and entered production in mid-1939 as the Sdkfz 251 mittlerer Schützenpanzerwagen (mSPW) medium infantry armoured vehicle. Some 68 Sdkfz 251 halftracks were in service by the outbreak of the war, and they proved a useful supplement to the predominantly truck-borne rifle infantry of Germany's panzer divisions. Inevitably the Sdkfz 251 reflected a series of compromises. At the time of its development it was not intended to be a direct fighting vehicle, but simply a transporter of infantry to the edge of the battlefield, where they would then disembark. The vehicle was thus relatively lightly armoured in order to maintain speed and cross-country performance to keep up with fast-moving panzer forces.

Production of the Sdkfz 251 steadily increased from 348 during 1940 to peak at 7800 during 1944, and by the end of the war over 16,000 had been built. As the war progressed, the basic vehicle underwent four major design simplifications, all intended to boost construction. The first variant, the Model A, appeared in 1939 and was distinguishable by its three prominent vision ports on each side of the superstructure. The Model B entered production in 1940 with the side vision ports omitted and a new distinctive shield fitted for the forward MG 34. In mid-1940 the Model C entered service. This version featured a single front plate fitted on the nose instead of the angled two-piece of earlier models. Finally, in 1942, the entire design was simplified to produce the Model D, which incorporated single large plates to replace the earlier faceted ones. This final Sdkfz 251 variant remained in production essentially unaltered until 1945.

The Waffen-SS acquired the halftrack much later than the German Army. It was not until 1942 – when Hitler decided to upgrade the *Leibstandarte*, *Das Reich* and *Totenkopf* to full panzergrenadier divisions in reward for their sterling service in the East – that the Waffen-SS began to receive halftracked APCs in any numbers. The Sdkfz 251 carried a complete 10-man Waffen-SS rifle squad plus their machine gun. Limited production, however, ensured that only about 160 Sdkfz APCs could be issued to each panzer division, sufficient only to equip a single panzergrenadier battalion. Although the Waffen-SS received priority in the receipt of the Sdkfz 251, there were still never enough to go round. Indeed, the number of these halftracks in German service peaked at 6146 vehicles on 1 December 1944. On this date, for example, the five élite SS panzer divisions refitting for the Ardennes Offensive in the West – the *Leibstandarte*, *Das Reich*,

Hohenstaufen, *Frundsberg* and *Hitlerjugend* Divisions – deployed 702 of their 785 authorised Sdkfz 251 halftracks.

By the time the Waffen-SS acquired the Sdkfz 251 its tactical employment had already began to change. In 1939 the Germans envisaged the Sdkfz 251 simply as a troop transporter that would deliver and drop riflemen at the edge of the bat-

tlefield, as mentioned above, where the grenadiers would enter combat before remounting. But during the 1940 Western Campaign, half-track-mounted panzergrenadiers were frequently in the thick of battle, moving forward alongside the armour and providing the latter with valuable support. Against Germany's ill-prepared early-war opponents who simply could not cope with

Sdkfz 234/2

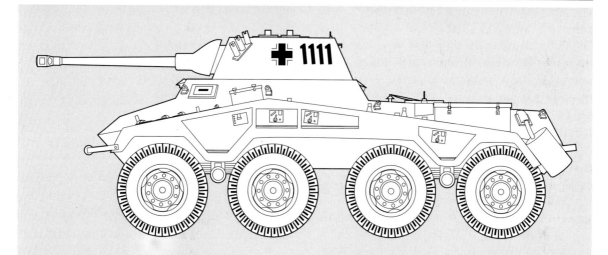

German Designation: Sdkfz 234/2 schwere Panzerspahwagen (sPSW) (5cm) Puma
Vehicle Type: Heavy eight-wheeled Armoured Car
Crew: four
Weight: 11.7 tonnes (11.5 tons)
Overall Length: 6.8m (22ft 4in)
Hull Length: 6m (19ft 8in)
Width: 2.4m (7ft 10in)
Height: 2.3m (7ft 6in)
Main Armament: 1 x 50mm KwK 39/1 L/60
Ammunition Stowage: Main = 220 rounds; Secondary = 2000 rounds
Armour:
 -Hull Front (Nose): 30mm (1.2in) (at 35 degrees)
 -Hull Front (Driver's Plate): 30mm (1.2in) (at 55 degrees)
 -Hull Sides: 9mm (0.4in) (at 60 degrees)
 -Hull Rear: 13mm (0.5in) (at 50-75 degrees)
 -Turret Front: 30mm (1.2in) (at 70 degrees)
 -Turret Sides: 10mm (0.4in) (at 65 degrees)
 -Turret Rear: 10mm (0.4in) (at 65 degrees)
 -Turret Roof: 10mm (0.3in) (at 0 degrees)
Engine: 220 bhp Tatra 103 12-cylinder air-cooled diesel
Fuel capacity: 405 litres (89 gallons)
Maximum Road Speed: 85kph (53mph)
Maximum Cross-Country Speed: 30kph (18.7mph)
Operational Range (Road): 1000km (620 miles)
Operational Range (Cross-Country): 805km (500 miles)

the speed at which the Germans conducted mobile warfare and the shock action this tempo inflicted, the Sdkfz 251 could more than hold its own on the battlefield. In France, therefore, German halftracks found themselves accompanying armour, screening the flanks of spearheads and de-bussing their troops in the middle of the battlefield to mop up a bewildered enemy scattered by rampaging German armour.

HALFTRACK TACTICS

The successes in France ensured that by the onset of Operation 'Barbarossa' in June 1941 German offensive doctrine had afforded halftracks a frontline combat role, which both increased their vulnerability and created a demand for better-armed and armoured halftracks to provide fire support. This latter need led ultimately to the emergence of no less than 22 variants of the basic Sdkfz 251 halftrack! The first variant was the Sdkfz 251/1, which carried a quadruple mount for a heavy 7.92mm MG 34 machine gun on its adjustable tripod mount and fitted with its long-range optical sight. On this version the armoured shield was omitted and a second heavy machine gun carried in the vehicle. This variant was issued to Waffen-SS heavy weapons platoons. Some Sdkfz 251/1, attached to SS armoured engineer platoons, were further modified to fire six 28cm/32cm Wurfkorper rockets. Launcher frames were fitted along each side of the vehicle at an elevation of 45 degrees. To fire, the entire vehicle had to be slewed onto the target and fired to a fixed range. Nevertheless, the 'Ground Stuka' as troops dubbed the vehicle delivered considerable firepower and it proved a cheap and quick-to-produce improvisation. The rockets, however, had to be fired in succession, as the back blast from a full salvo could overwhelm the crew and even overturn the vehicle!

The Germans also adapted the Sdkfz 251 to carry the standard 8.1cm mortar of the heavy weapons company of the panzergrenadier battal-

ion. A few minor modifications even allowed the mortar to be fired directly from the vehicle. The Sdkfz 251/5 Pioneerpanzerwagen – armoured engineer vehicle – possessed side racks to allow pontoon bridge sections or assault boats to be carried externally. The interior was also modified to allow greater storage capacity for various engineer stores, such as mines and demolition charges. This vehicle was issued to armoured engineer companies of Waffen-SS panzer divisions. The Sdkfz 251/6 Kommandopanzerwagen was a rare, specialised command vehicle intended for senior commanders, and thus only produced in limited quantities. It incorporated office facilities, including a folding map table, cypher and encoding devices and several radios, and these features made it the preferred command

vehicle for most senior Waffen-SS field officers
during the latter half of the war.

GREATER ARMAMENT

The increasing combative role of halftracks
ensured that German firms incorporated into the
vehicle a heavier main armament to give it greater
offensive punch. The Sdkfz 251/10 was the first
such variant and carried a 3.7cm Pak 36 anti-tank
gun. The entire gun and shield was simply mounted on the forward superstructure. It first entered
service in 1941, and field experience soon
revealed that the full shield was a hindrance; later
production models, therefore, carried either a
smaller shield, a half-shield or no shield at all.

The second gun-armed variant was the Sdkfz
251/9 introduced in 1942. This carried the short-

barrelled 7.5cm (2.95in) KwK 37 L/24-calibre gun
formerly mounted in the Panzer IV tank. This
vehicle gave Waffen-SS panzergrenadiers their
own indigenous fire-support capability. The cannon company of SS panzergrenadier regiments
received six of these vehicles, affectionately
dubbed by the troops that operated them as 'the
Stump' (Stummel). Another potent variant was
the Sdkfz 251/16 Flammpanzerwagen which carried two 700-litre (154-gallon) flame-fuel tanks
and two 14mm (0.55in) flame projectors mounted well back on the superstructure top on both
sides of the vehicle. Each projector could tra-

verse through 160 degrees and some vehicles possessed a third portable flame projector, the modified Flammenwerfer 41 with cartridge ignition projector, on a long 10.1m (33ft) extension hose for use outside the vehicle. Since the fuel tanks were housed in the rear of the passenger compartment, the rear doors had to be welded shut. The vehicle could fire 80 two-second bursts of flame to a distance of 35m (38yds). From late 1943, six Waffen-SS armoured engineer companies each received six of these vehicles.

BELOW: *A late-production Sdkfz 250/3 armoured command halftrack on the Eastern Front in October 1943, complete with a pole and star aerial.*

The growing Allied air threat led to the introduction of the Sdkfz 251/17 in 1943. This variant mounted a 2cm Flak 30 or Flak 38 on top of the rear compartment and fired from within. Some possessed modified, hinged bulging sides that could be lowered for improved traverse. A few examples produced from late 1944 had the gun mounted in an enclosed remote-controlled turret. By 1944, Allied aerial mastery had become so great that the Germans introduced the Sdkfz 251/21 Flakpanzerwagen which mounted three surplus Luftwaffe 15mm MG 151 machine guns in a triple-shielded co-axial mount. Though each gun could fire up to 700 rounds per minute, the vehicle carried only 3000 rounds of ammunition.

The conversion produced a cheap but effective addition to the anti-aircraft artillery companies of select SS panzergrenadier regiments during the last year of the war.

By 1944, the ever-more desperate need for firepower in the face of increasing numbers of enemy armour led to the introduction of a final variant, the Sdkfz 251/22, which carried the 7.5cm Pak 40 anti-tank gun mounted on the fighting compartment of the halftrack. After removal of the wheels and trails of the gun, as well as the roof of the driving cab, the gun still only had very limited traverse. Nonetheless, this conversion proved an effective expedient that considerably enhanced the firepower of Waffen-SS panzergrenadiers, though it was only produced in small numbers.

Alongside the Sdkfz 251 served the Sdkfz 250 leichter Schützenpanzerwagen. In 1939 the High Command concluded that the Demag Sdkfz 10 artillery tractor was suitable as a basis for a smaller, light halftracked APC. The Sdkfz 250 featured a scaled-down version of the armoured body of the Sdkfz 251 and could only carry six men. Nonetheless, the smaller vehicle proved better suited for a variety of specialised roles, particularly as a command vehicle or as a mortar carrier. In total, German firms constructed 14 different models of the Sdkfz 250.

UPGRADING THE SDKFZ 250

The standard Sdkfz 250 had a combat weight of 5.3 tonnes (5.2 tons), possessed a 1-tonne (0.98-ton) payload and could reach a top speed of 60kph (37mph). The vehicle featured the same protection as its larger cousin: 14.5mm (0.6in) armour on the front and 8mm (0.3in) on the rear and sides. The Sdkfz 250 entered service in early 1940 and first saw combat in the May 1940 Western Campaign. From 1943 the vehicle underwent repeated simplifications to speed production, such as the building-in of stowage lockers, replacement of vision flaps with slits and

enlargement of the rear door to improve egress. Since the basic troop carrier, the Sdkfz 250/1, could not carry a full infantry squad, it was often used as a platoon and company commander's vehicle. Like its larger sister, the Sdkfz 250 usually carried two 7.92mm MG 34 machine guns on pivot mounts on the superstructure front and rear.

SPECIALISED VARIANTS

As the war progressed, the Sdkfz 250 found itself increasingly in the midst of combat at the front, and consequently the Germans developed a range of variants to fulfil a variety of specialised roles. Indeed, the basic Sdkfz 250/1 was usually only issued to armoured reconnaissance and engineer companies. One such variant was the Sdkfz 250/3 leichter Funkpanzerwagen radio vehicle that carried the tell-tale frame aerial of a command halftrack. General Erwin Rommel used one of these halftracks, named Greif (Griffin), as his personal command vehicle in North Africa during 1942. Later Sdkfz 250/3 models featured less cumbersome pole and star aerials in place of the frame aerial.

The Sdkfz 250/7 was a dedicated mortar carrier that increasingly replaced the Sdkfz 251/1. The mortar could be fired from the vehicle, but the squad preferred to deploy the mortar and fire from covered ground. The Sdkfz 250/8, which entered service in 1943, carried the 7.5cm KwK L/24 gun. On this model a 7.92mm MG 42 machine gun was mounted above the main armament and provided with tracer ammunition to allow it to act as a sighting/ranging device for the main armament, as well as for close defence. Six of these vehicles equipped the heavy gun platoon of select Waffen-SS panzergrenadier battalions.

When German armoured cars proved unable to withstand the harsh conditions of the war on the Eastern Front, the German High Command decided to replace these vehicles with halftracks wherever possible since the latter enjoyed better survivability, mobility and mechanical reliability

ABOVE: *A Waffen-SS Sdkfz halftrack armoured personnel carrier passes a Panzer III tank during a German assault on the Eastern Front.*

than wheeled vehicles. The result was a need for better-armed halftracks, and so the Sdkfz 250/9 was designed as a replacement for the Sdkfz 222 light armoured car. The halftrack simply mounted the entire turret of the Sdkfz 222 armoured car atop the superstructure of the halftrack, which was roofed-in.

Another self-propelled gun variant was the Sdkfz 250/10, which mounted a 3.7cm Pak 36 anti-tank gun. The need for increased anti-tank capability also resulted in the Sdkfz 250/11 in 1942. This vehicle carried the 2.8cm schwere Panzerbüsche 41 taper-bore anti-tank rifle mounted on the superstructure front. The Sdkfz 250, however, was manufactured in smaller numbers than the Sdkfz 251 and consequently

remained rarer in Waffen-SS service. On 1 November 1944 there were just 2185 Sdkfz 250 APCs in the entire German armed forces. The five SS panzer divisions then refitting in the West, for example, were each authorised just 34 of these light APCs, and together they had 174 on strength. The vehicle continued to offer dedicated service until the German capitulation on 8 May 1945.

DEPLOYMENT WITHIN SS DIVISIONS
From 1942 the premier SS divisions begun to receive the Sdkfz 250 and 251 in numbers for the first time, and the latter became the standard transporter in their panzergrenadier regiments. Production shortages, however, meant that only one panzergrenadier battalion in each SS panzer division received halftracks, this unit being designated 'armoured' (gepanzerte). Each of its 183-man panzergrenadier companies deployed 12

Sdkfz 251

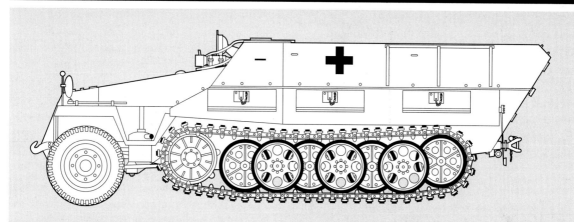

German Designation: Sdkfz 251 mittlerer
  Schützenpanzerwagen (mSPW)
Vehicle Type: Medium Armoured Personnel Carrier
Crew: 12
Weight: 7.8 tonnes (7.7 tons)
Length: 5.80m (19ft)
Width: 2.10m (6ft 11in)
Height: 1.75m (5ft 9in)
Armament: 2 x 7.92mm (0.312in) MG 34 machine
  guns
Ammunition Stowage: 2000 rounds

Armour:
  -Front (Nose): 14.5mm (0.6in)
  -Front (Driver's Plate): 14.5mm (0.6in)
  -Sides: 8mm (0.3in)
  -Rear: 8mm (0.3in)
Engine: 100 bhp Maybach NL 42 six-cylinder petrol
Fuel capacity: 200 litres (44 gallons)
Maximum Road Speed: 52.5kph (32.5mph)
Maximum Cross-Country Speed: 29kph (18mph)
Operational Range (Road): 300km (186 miles)
Operational Range (Cross-Country): 130km
  (80 miles)

halftracks organised in three platoons, plus a heavy weapons platoon of two Sdkfz 251/2 mortar carriers and two 7.5cm-gunned Sdkfz 251/8 halftracks.

Waffen-SS employment of the armoured halftrack evolved as the war progressed. Initially, APC-mounted SS panzergrenadiers scouted ahead of armour and deployed on the flanks to guard against enemy counterattacks as the panzers punched through thin enemy defences. As Allied firepower increased and the Germans lost the strategic initiative during the middle of the war, SS panzergrenadiers gradually changed tactics, however. They increasingly abandoned their

previous flank protection role – which had become too dangerous and costly – and instead often preceded the armour to locate and neutralise enemy anti-tank gun nests and disperse enemy infantry tank-destruction parties waiting to ambush armour at close range.

During 1944, instead of SS panzergrenadiers supporting armour, the tanks increasingly supported the panzergrenadiers as the halftrack proved steadily more vulnerable to enemy fire. Ironically, the defensive employment of the halftracks ultimately turned full circle: they reverted to their original role of transporting troops to the very forward edge of the battlefield, where the panzergrenadiers de-bussed prior to engaging the enemy from fixed defences while the weapons-armed halftracks provided fire support from the relative safety of rear and flank positions. Offensives, on the other hand, still required panzergrenadiers to spearhead the attack, clearing ground of enemy infantry and anti-tank guns to allow the armour to advance. Indeed, continued offensive and defensive success demanded intimate combination of arms – today called the all-arms battle.

BELOW: *A factory new Sdkfz 251/7 Panzerpioneer-wagen armoured engineer vehicle. Note the bridge-building material carried on its sides.*

By 1945 SS panzergrenadiers thus fought with tactics very different from those they had used early in the war. Now, small mobile groups of SS tanks and panzergrenadiers were constantly 'fire brigaded' and moved from one danger spot to another. Their mission was to halt enemy penetrations and buttress an increasingly fragile German front. They only gave ground grudgingly and utilised their firepower to exact a heavy price for every metre conceded.

The Waffen-SS anti-aircraft artillery units employed a variety of light and medium self-propelled flak guns that were mounted on half-tracked artillery tractors. Early in the war, the Germans recognised that the artillery tractors

that had spawned the Sdkfz 250 and 251 APCs could also mount a variety of flak guns to provide the Waffen-SS with its first self-propelled anti-aircraft artillery.

The lightest tractor was the 4.9-tonne (4.8-ton) Sdkfz 10, which the Germans used both as an eight-man troop transporter and to tow payloads of up to one tonne (0.98 ton). Entering service in 1937, its 100bhp Maybach HL 42 TRM petrol engine produced a top speed of 65kph (40mph) and a radius of 300km (186 miles) over roads and 170km (106 miles) cross-country. It was the most widely manufactured German prime mover, with over 17,000 constructed.

FLAK HALFTRACKS
In 1939, the 2cm Flak 30 and Flak 38 guns were mounted on the light Sdkfz 10 artillery tractor to produce the Sdkfz 10/4 and 10/5 respectively. The addition of these weapons raised the weight of both vehicles to 5.6 tonnes (5.5 tons). Hinged side and rear railings could be dropped down to provide a wider firing platform for the gunners when the vehicle was in action against either aerial or ground targets. Some vehicles also featured an armoured cab fitted over the formerly open driver's compartment to provide protection for the crew. Waffen-SS field workshops also improvised a number of unofficial marriages between the Sdkfz 10 light artillery tractor and available guns, such as mounting the entire 3.7cm Pak 35/36 anti-tank gun with shield on the rear of the tractor facing forward, so that the gun fired over the cab.

Later in the war the 2cm Flakvierling 38 quadruple gun was mounted on the Krauss-Maffei Sdkfz 7 halftrack to create the Sdkfz 7/1. In its basic configuration the Sdkfz 7 weighed 11.2 tonnes (11 tons) and was driven by a 140bhp Maybach water-cooled engine that provided a top road speed of 50kph (31mph). The vehicle's operational road range was 250km (155 miles) and 120km (74.5 miles) cross-country. On the Sdkfz 7/1 the entire 20mm quad flak mount was simply bolted onto the rear fighting compartment. On many of these vehicles the driver's cab was enclosed and armoured for greater protection. This self-propelled gun was issued mainly to the anti-aircraft companies of SS panzergrenadier regiments and provided column cover against low-flying aircraft.

DAS REICH FLAK UNITS IN ACTION
Typical of the employment of this vehicle was the experience of the 2nd SS Panzer Division *Das Reich* as it proceeded toward the Normandy beaches during mid-June 1944. Its columns of Panther tanks advanced separated by 50m (55yd) intervals, and the division interspersed Sdkfz 7/1 vehicles between each platoon to provide cover against low-flying Allied aircraft. Sure enough, near Periers American Thunderbolt fighter-bombers strafed the column and came round for a second pass. The nearest Sdkfz 7/1 self-propelled flak gun then roared into action. Its crew drove the vehicle off the side of the road, threw down its side rails, cleared the guns for action and opened up with its four barrels, spewing forth hundreds of 2cm rounds that quickly caught the lead Thunderbolt as it flew over the column. As this hail of fire punched through the aircraft's fuselage, the Thunderbolt first veered off toward the north trailing plumes of black smoke, and then plunged down to crash into nearby woods. The other American fighter-bombers abandoned their strafing runs in the face of such heavy defensive fire and departed in search of easier targets, permitting the SS column to continue its journey to the invasion front. The Sdkfz 7/1 had won this particular duel, but the scenario would be repeated many more times during the Normandy campaign – and often the outcome proved less fortunate for the SS troops.

The success of the 2cm-equipped self-propelled flak halftrack early in the war soon led to the mounting of heavier guns on artillery tractors. The 3.7cm Flak 36 gun, for example, was ultimately mounted on two tractors. The first vehicle – based on the standard Bussing-NAG Sdkfz 6 tractor – weighed 9.8 tonnes (9.8 tons) and was powered by a 115bhp Maybach petrol engine that produced a maximum speed of 50kph (31mph) and an operational range of 310km (193 miles). This flak halftrack, designated the Sdkfz 6/2, first entered service in late 1939 and mounted the 3.7cm gun on an open platform behind the driver's cab. As in previous models, the wooden side and rear rails folded down to provide a wider platform for the crew. The second vehicle – the Sdkfz 7/2 – likewise mounted a single 3.7cm Flak 36 on the back of the standard Sdkfz 7 medium tractor. These vehicles served in Waffen-SS self-propelled medium flak batteries throughout the second half of the war.

WAFFEN-SS SELF-PROPELLED FLAK GUNS

The Waffen-SS first began to receive self-propelled flak guns during 1941, and the number in service steadily increased thereafter. By 1944 SS panzer divisions had become well equipped with anti-aircraft guns. They had on strength a flak battalion that included several light and medium self-propelled batteries. In addition, divisional heavy motorised flak batteries often received a light self-propelled platoon of three halftracks to protect their 8.8cm Flak 41 guns from air attack. The premier SS panzer and panzergrenadier divisions also possessed a light self-propelled platoon in their escort companies that usually fielded the Sdkfz 7/1 quadruple flak mount. Waffen-SS panzergrenadier regiments also typically fielded a self-propelled company equipped with 12 2cm-gunned Sdkfz 10/4 and 10/5 vehicles. Finally, panzer battalion headquarters received a light self-propelled flak platoon equipped with another six of these vehicles.

During 1944 many SS grenadier divisions also received a motorised anti-aircraft company that consisted of either 12 2cm light or nine 3.7cm medium flak guns. Most divisions received only towed weapons, but several divisions, including the 14th and 15th SS Volunteer Grenadier Divisions (Galician No 1) and (Latvian No 1) respectively, fielded companies that were at least partially equipped with flak halftracks. The rarest Waffen-SS self-propelled flak vehicles were, however, the Sdkfz 7/2 that mounted the quad 2cm Flakvierling gun. These generally served only in the headquarters companies of SS panzer artillery regiments, which usually

deployed a platoon of three or four Sdkfz 7/2 tractors. The vehicle was also found at corps level – the I, II and IV SS Panzer Corps, for example, each fielded two platoons of three Sdkfz 7/2 self-propelled flak vehicles.

Armoured cars, APCs and halftracked self-propelled flak guns performed a variety of specialised and very important missions and ultimately became an integral element of the Waffen-SS armoury during World War II. They buttressed both the offensive and defensive capabilities of SS troops and helped to ensure that the Waffen-SS gained and maintained its reputation as a formidable and effective military élite. More often than not, they operated on the battlefield in intimate cooperation with Waffen-SS tanks, artillery and self-propelled guns, and thus it is to the story of Germany's early war panzers and their use in the Waffen-SS that the following chapter turns.

BELOW: *Officers of the* Leibstandarte *next to a Sdkfz 251/10 with a 3.7cm anti-tank gun, which belongs to the 1st SS Armoured Reconnaissance Battalion.*

CHAPTER 5

Light and Medium Tanks

It was with Germany's first generation of pre-war tanks, the Panzer II, III and IV, that the Waffen-SS cut its teeth in the giant armoured steppes of Russia. Despite being outnumbered and often out-gunned, the Waffen-SS Panzer IIIs and IVs gave sterling service.

The weapon most associated with the Waffen-SS during World War II is the tank, or panzer. Although the Waffen-SS only first acquired significant amounts of armour during 1942, for the rest of the war the tanks deployed by the élite Waffen-SS panzer divisions made an enormous contribution to Germany's defensive achievements on all fronts. This chapter describes the first generation pre-war tanks utilised by the wartime SS – the Panzer II, III and IV – as well as various captured foreign tanks. Of these, only the Panzer IV remained in frontline SS service until the end of the war, by which time even it had become inferior to the latest Allied tanks and

LEFT: *A Panzer IV Model H of the* Hohenstaufen *at Lisieux, France, in June 1944. Note the Zimmerit anti-magnetic mine coating on all external surfaces.*

anti-tank weapons. Instead, during the last two years of the war, the Waffen-SS increasingly relied on a second generation of tanks: the Panzer V Panther, the Panzer VI Tiger I and the Panzer VIB King Tiger.

The Panzer I, the first light training tank developed by Nazi Germany after 1933, never served with the Waffen-SS since it had already become obsolete long before the Waffen-SS received tanks in numbers for the first time. The oldest tank that the Waffen-SS did employ during World War II was the Panzer II. In 1934, the German Army ordered a light tank as a stop-gap until its main combat tanks – the Panzer III and IV – entered operational service. German firms commenced mass production of the Panzer II Model A tank in 1937. This vehicle weighed 8.9 tonnes (8.8 tons), mounted as its main armament

ABOVE: *A Panzer III Model H armed with a short-barrelled 42-calibre 50mm gun of the* Totenkopf *Division in Russia during October 1942.*

a 2cm KwK 30 L/55 gun and possessed 14.5mm-(0.6in-) thick armour. It was powered by a 140bhp Maybach HL 62 TRM engine, and its suspension comprised five large road wheels hung from quarter-elliptic springs. During 1937–39, German firms produced the similar Models B and C, as well as the Model D and E fast-reconnaissance tanks that were capable of 56kph (35mph) by road. Manufacture commenced in March 1941 of the 9.5-tonne (9.3-ton) Panzer II Model F, which featured a homogenous 35mm- (1.4in-) thick frontal plate, and side armour increased to 30mm (1.2in). During 1942–43, the Model F was also the first variant of the Panzer II tank to serve with the Waffen-SS.

The experiences the Germans gained during Operation 'Barbarossa' demonstrated that the Panzer II was so under-gunned and under-armoured that it could not fight effectively on the Eastern Front. The Germans, however, contin-

ued to produce Panzer II Model F tanks – and the latest G and J models – until late 1942 despite their growing obsolescence. In the East during the winter of 1942–43, the Waffen-SS increasingly utilised the Panzer II as a reconnaissance vehicle. From early 1943, however, the High Command gradually withdrew the Panzer II from frontline service, and converted many of them into Marder light tank destroyers and Wespe self-propelled howitzers.

The Panzergrenadier Division *Leibstandarte* was the only frontline Waffen-SS formation to employ the Panzer II Model F tank. In late 1942, the 5th Company of its new tank regiment received 12 Panzer II Model F vehicles manufactured earlier that summer. In January 1943, the Germans redeployed the *Leibstandarte* to the Eastern Front as part of I SS Panzer Corps. Here, its dozen Panzer II tanks performed scouting missions during the bitter fighting that raged around Kharkov during the spring. By 4 March the *Leibstandarte* had lost eight of its Panzer II tanks, and in April withdrew the remaining four vehicles from frontline combat. This decision

Panzer II Model F

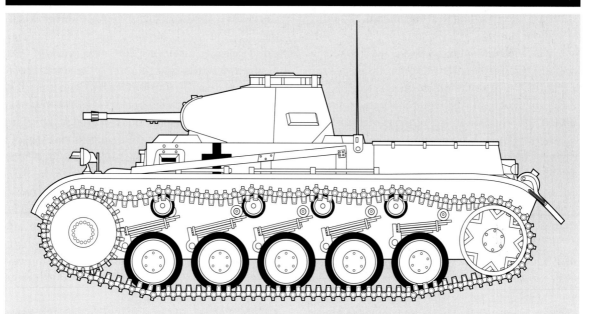

German Designation: Panzer II Model F (Sdkfz 121)

Vehicle Type: Light Tank

Crew: Three

Weight: 9.5 tonnes (9.3 tons)

Overall Length: 4.81m (15ft 9in)

Hull length: 4.81m (15ft 9in)

Width: 2.28m (7ft 6in)

Height: 2.02m (6ft 7in)

Main Armament: 2cm (0.8in) KwK 30 (or KwK 38) L/55 gun

Secondary Armament: 2 x 7.92mm (0.312in) MG 34; 1 co-axial in turret; 1 hull front.

Ammunition Stowage: Main = 180 rounds; Secondary = 2550 rounds

Armour

-Hull Front (Nose): 35mm (1.4in) (at 77 degrees)

-Hull Front (Driver's Plate): 30mm (1.2in) (80 degrees)

-Hull Sides: 20mm (0.8in) (at 90 degrees)

-Hull Rear: 14.5mm (0.6in) (at 90 degrees)

-Turret Front: 30mm (1.2in) (at 90 degrees)

-Turret Sides: 15mm (0.6in) (at 68 degrees)

-Turret Rear: 14.5mm (0.6in) (at 68 degrees)

-Turret Roof: 10mm (0.4in) (at 0-13 degrees)

Engine: Maybach HL 62 TRM R six-cylinder petrol

Power: 140bhp

Fuel capacity: 170 litres (30.81 gallons)

Maximum Road Speed: 40kph (25mph)

Maximum Cross-Country Speed: 19kph (12mph)

Operational Range (Road): 200km (124 miles)

Operational Range (Cross-Country): 130km (81 miles)

implemented an earlier policy to withdraw the remaining 300 Panzer II tanks still in frontline service. Consequently, by the spring of 1943 the brief service of the Panzer II with the Waffen-SS had come to an end.

Germany's first medium tank, the Panzer III (Sdkfz 141) which entered service in 1936, mounted the 3.7cm KwK L/46.5. The first six variants produced prior to July 1940 – the Models A-F – did not serve with regular Waffen-SS formations, because the Germans had withdrawn these vehicles from frontline service long before the Waffen-SS received its first significant allocation of armour during 1942. For the remainder of the war the Germans utilised these aged tanks as training vehicles for new recruits, and in this role a few briefly entered Waffen-SS service. In April 1943, for example, the 9th SS Panzer Division *Hohenstaufen*, then in the process of forming in northern France, received two 1936-vintage Panzer III Model A tanks – only 10 had ever been produced – with which to train the division's green recruits. The division relinquished these tanks when it transferred to the Eastern Front during the spring of 1944.

In April 1940, the Panzer III Model G armed with a more powerful 5cm KwK L/42 gun entered

BELOW: *Waffen-SS armour of I SS Panzer Corps at Kharkov in early 1943. The last tank appears to be a Panzer III Model L.*

service and some 450 vehicles were produced up to February 1941. Between October 1940 and April 1941, German firms then produced a further 310 Panzer III Model H tanks that incorporated thicker armour. The next variant, the Model J, was the first to enter widespread frontline service with the Waffen-SS. The first few hundred Model J tanks constructed still retained the 5cm KwK L/42 gun, but featured 50mm (2in) homogenous frontal armour. Subsequent Model J tanks – designated Sdkfz 141/1 – mounted the more potent long-barrelled 5cm KwK 39 L/60 gun, which increased the vehicle's weight to 22.3 tonnes (21.9 tons). German factories constructed 2516 Model J tanks between March 1941 and July 1942. This variant equipped the panzer battalions that the Germans added to the SS Motorised Divisions *Das Reich* and *Wiking* during the spring and summer of 1942.

The Panzer III Model L featured the same powerful long 5cm KwK L/60 gun as its prede-

Bergepanzer III

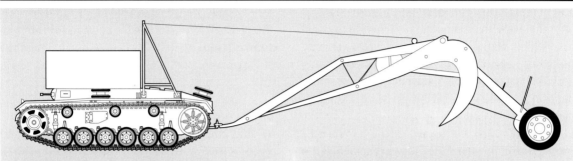

German Designation: Bergepanzer III
Vehicle Type: Armoured Recovery Vehicle
Crew: Three
Weight: 19 tonnes (18.7 tons)
Overall Length: 6.28m (20ft 7in)
Hull length: 6.28m (20ft 7in)
Width: 2.95m (9ft 8in)
Height: 2.45m (8ft)
Armament: 2 x 7.92mm (0.312in) MG 34
Ammunition Stowage: 1000 rounds
Armour:
 -Hull Front (Nose): 50mm (2in) (at 69 degrees)
 -Hull Front (Driver's Plate): 50mm (2in) (at 81 degrees)

 -Hull Sides: 30mm (1.2in) (at 90 degrees)
 -Hull Rear: 30mm (1.2in) (at 77-80 degrees)
 -Superstructure Front: Wood (at 90 degrees)
 -Superstructure Sides: Wood (at 90 degrees)
 -Superstructure Rear: Wood (at 90 degrees)
 -Superstructure Roof: Wood (at 0 degrees)
Engine: 300 bhp Maybach HL 120 TRM V-12 cylinder petrol
Fuel capacity: 320 litres (70 gallons)
Maximum Road Speed: 40kph (25mph)
Maximum Cross-Country Speed: 25kph (15.5mph)
Operational Range (Road): 200km (124 miles)
Operational Range (Cross-Country): 110km (68 miles)

cessor, but also incorporated more effective spaced armour. German factories produced 703 of these tanks between June and December 1942. The Model L equipped the panzer battalions raised during the summer of 1942 for the *Leibstandarte* and *Totenkopf* Divisions. The Panzer III Model M, which entered service in late 1942, incorporated minor modifications such as thin armoured side-skirts (Schürzen) to protect its wheels and tracks from hollow-charge anti-tank weapons. German firms constructed 292 Model M tanks between October 1942 and February 1943. This variant reached Waffen-SS forces in early 1943, in time to participate in the bitter battles fought by I SS Panzer Corps around Kharkov. Even before these battles, however, the

Germans had recognised the growing battlefield inferiority of even the latest Panzer III tanks. But despite this, the need to produce the maximum number of vehicles to offset heavy losses meant that the Germans increased Panzer III production during 1942 with the Models J-M. Although Model M tanks featured prominently in the ranks of I SS Panzer Corps at Kursk in July 1943, this battle was the last great armoured encounter in which the Panzer III participated in large numbers.

The Model M represented the last standard, multi-purpose Panzer III to be introduced. The last Panzer III type, the Model N (Sdkfz 141/2), was designed for the heavy close fire-support role, and soon earned the popular name Sturmpanzer III assault tank. The Model N mounted the short-barrelled 7.5cm KwK L/24 gun previously mounted in Panzer IV tanks prior to March 1942. This gun delivered a poor armour-piercing capability, but a good high-explosive one which suited its designated role. German factories constructed 666 Model N tanks between late 1942 and August 1943. A mere handful of these vehicles served with Waffen-SS panzer divisions during 1943.

In August 1943, the Germans halted Panzer III production in favour of the Panzer IV tank and the cheaper StuG III assault gun, after a total production run of 5623 battle tanks. This figure included 673 Models A-F that mounted the 3.7cm KwK L/46.5 gun; 2315 Models G, H and J with the short 5cm L/42 weapon; 1969 Model J, L, and M tanks armed with the long 5cm L/60; and 666 Model N tanks mounting the short 7.5cm KwK L/24 gun.

THE RUGGED PANZER III

The Waffen-SS also employed small quantities of the Panzerbefehlswagen III Model H command tank. This was a modified Panzer III tank with a dummy gun and prominent rail antenna on its rear hull. German factories produced 175 Model H command tanks between November 1940 and January 1942. The Waffen-SS also fielded some of the 235 Panzer III Model J, K and M command tanks constructed or converted between August 1942 and February 1943. These vehicles represented an improvement on their predecessors as they retained their main armament. Small numbers of these Model J-M command tanks remained in frontline SS service until the spring of 1945.

The Panzer III displayed its ruggedness and versatility during Operation 'Barbarossa'. The Germans deployed 965 Panzer III tanks for the invasion, but none of these served with the Waffen-SS, since its motorised infantry divisions did not possess armour. The short 5cm-armed Panzer III Models G, H and early J, together with the few long-barrelled Model J tanks that took part in the invasion, made a critical contribution

to the enormous German successes achieved in the opening stages of 'Barbarossa'. The sterling service the Panzer III provided ensured that when the Waffen-SS finally acquired armour in numbers during 1942, it received large numbers of Panzer III tanks.

THE CHALLENGE OF RUSSIA'S TANKS

But as soon as the Germans engaged the T-34 medium and KV-1 heavy tanks – the newest and most formidable Soviet armour – they soon recognised the inferiority of the Panzer III tank. Indeed, the T-34 mounted a 7.62cm gun at a time when the Germans were just up-gunning the Panzer III from a 3.7cm to a 5cm gun. Both the 90mm (3.6in) armour of the KV-1 and the well-sloped 45mm (1.8in) plates of the T-34, rendered them virtually immune to fire from both the German 3.7cm and 5cm tank guns at typical combat ranges. At Mtsensk, near Orel, on 4 October 1941, for example, the German Army's 4th Panzer Division had just secured a bridgehead over the River Lisiza at Kamenewa when the Red Army counterattacked. Some 50 Soviet T-34 and KV-1 tanks ambushed the German force and destroyed 10 Panzer III tanks. To their horror, the Germans discovered that rounds fired by the 5cm KwK L/42 guns of their Model G and H Panzer III tanks failed to penetrate the frontal armour of the KV-1 at ranges as short as 800m (875yds)!

As these new Soviet tanks began to contribute towards the blunting of the German Blitzkrieg during 1942, the German High Command converted the premier SS divisions – *Leibstandarte, Das Reich, Totenkopf* and *Wiking* – from motorised formations to panzergrenadier ones. This process involved the initial addition of a three- or four-company panzer battalion during the summer of 1942 that fielded Panzer III and IV tanks. Subsequently, during the winter of 1942–43 these battalions were expanded to a two-battalion panzer regiment. During April 1942, while *Wiking* remained on the Eastern

ABOVE: *A column of Waffen-SS Panzer IV Model E tanks on the move in Russia. This variant entered service at the end of 1940.*

Eastern Front, its new panzer battalion formed at the Sennelager training camp in Germany. With its new production Panzer III Model J tanks, the battalion rejoined *Wiking* in May just as the division prepared to join the German Operation 'Blue' offensive of the summer of 1942 that sought to capture the oil-rich Caucasus region.

During the summer of 1942, while replenishing in the west, the SS Panzergrenadier Division *Das Reich* received its three-company panzer battalion that included the new Panzer III Model J. The SS Panzergrenadier Division *Leibstandarte* likewise raised a panzer battalion during the summer of 1942, including large numbers of the new Panzer III Model L tank. In January 1943 Hitler dispatched these two divisions, under command of the new I SS Panzer Corps, to the East to stem the successful Soviet counteroffensive in the south. In October 1942, the High Command also withdrew the *Totenkopf* Division from the Eastern Front, re-equipped it with a panzer battalion that fielded many Model L Panzer III tanks, and returned it to the East in early 1943 to join its two sister divisions in the recently arrived SS panzer corps.

Field Marshal von Manstein flung I SS Panzer Corps, bristling with Panzer III Model J and L tanks, into the fighting around Kharkov. During February and March 1943, in terrible conditions of ice and snow, the SS corps spearheaded Manstein's masterful counter-stroke. This caught the extended Soviet spearheads off balance, and managed to both recapture Kharkov and drive back the Red Army beyond the River Donets. In these battles the first Panzer III Model M tanks reached I SS Panzer Corps and some of these vehicles replaced the obsolete Panzer II tanks of the 5th Panzer Company, 1st SS Panzer Regiment. Despite stunning success, I SS Panzer Corps suffered heavy losses in the Kharkov battles. By 4 March 1943, for example, *Das Reich* fielded just 11 operational Panzer III tanks, having lost all of its Panzer IV tanks in these engagements.

During the lull that followed in April 1943, the three élite divisions of I SS Panzer Corps enjoyed

ABOVE: *A Leibstandarte* Panzer IV Model J tank with *Schürzen aprons and Zimmerit anti-magnetic mine paste. Note the spare track sections on the hull front.*

a brief rest before the German High Command ordered the corps to spearhead the German offensive at Kursk, Operation 'Citadel', that summer. In June, the divisions received reinforcements, including new Panzer III Model M tanks. On 4 July 1943, the start of 'Citadel', the SS corps deployed 422 tanks and assault guns, about 120 of which were Panzer III Model J-M tanks. *Das Reich* alone fielded 35 Panzer III tanks.

By the time of the German onslaught at Kursk, however, the Panzer IV had already overtaken the Panzer III as the principal weapon of Germany's armoured forces. Indeed, one of the last significant achievements of the Panzer III on the Eastern Front was the contribution that Model J-M tanks made to the defensive successes accomplished by the 5th SS Panzergrenadier

Division *Wiking* around Stalino during August 1943. In these battles *Wiking* also deployed at least one very rare Panzer III Model L tank that for some reason mounted the short 5cm KwK L/42 gun instead of the then standard longer L/60 weapon.

From the autumn of 1943, the Germans gradually began to phase out the Panzer III from frontline service in the East, as new Panther battalions joined the panzer divisions. This policy, when combined with continuing heavy attrition rates, caused Panzer III strengths on the Eastern Front to plummet during the autumn of 1943. Consequently, by October 1943 only five panzer divisions in the East still deployed more than one company of Panzer III battle tanks; and by late 1944 just 79 Panzer III combat tanks remained in frontline service on the Eastern Front.

Even though Panzer III battle tanks became increasingly scarce after late 1943, small numbers of command tanks remained on active service throughout 1944. A 1944-pattern SS panzer division theoretically included 10 command tanks: five 1942-vintage Panzerbefehlswagen III Model H, identifiable by their distinctive rail antenna, plus five of the newer Model J and M variants. Two of these command tanks each served with the staff company and with the four company headquarters troops of the division's Panzer IV-equipped tank battalion. Combat losses meant that by the autumn of 1944, however, the Waffen-SS retained only a handful of Panzerbefehlswagen III Model H command tanks alongside approximately 80 of the more modern Command Panzer III Model M vehicles. In Normandy during the summer of 1944, for example, the 12th SS Panzer Division *Hitlerjugend* fielded six Panzer III Model M command tanks. In the December 1944 Ardennes Offensive, the four SS panzer divisions committed each fielded two or three of these vehicles, at least one of which succumbed to Allied aircraft during the attack. Indeed, as late as 1 April 1945 the 10th SS Panzer Division *Frundsberg* still fielded a solitary operational Panzer III command tank during the bitter defensive engagements it fought in Pomerania.

SWANSONG OF THE PANZER III

By the autumn of 1944, however, the Germans had relegated virtually every remaining Panzer III battle tank to training units. It was in this capacity that a few Panzer III tanks once again saw brief combat during the last year of the war. One such scenario was the activation of the SS Panzer Brigade *Gross*, raised from panzer training units at the Seelager proving grounds near Riga, Latvia, in August 1944. The brigade initially fielded five old Panzer III tanks that fought against Soviet armour on the northern sector of the Eastern Front during the autumn of 1944. Continued Allied success during 1944 also encouraged further resistance attacks against the Nazis, which compelled the Germans to reinforce their anti-partisan forces with obsolete

tanks. In early 1945, for example, German security forces behind the frontline in northern Italy waged an increasingly desperate struggle to prevent the Italian Resistance paralysing German communications. One unit committed to this brutal campaign was the SS Panzer Company *Liguria*, equipped with 14 old 1942-vintage Panzer III tanks. The Germans periodically diverted the *Liguria* from its normal duty – guarding Benito Mussolini's Italian Fascist headquarters at Lake Salo – to undertake anti-partisan sweeps. But in late April 1945 the Higher SS Police Leader in Italy, SS-Obergruppenführer (General) Karl Wolff, withdrew the *Liguria* from either of these duties to protect his own headquarters while he thrashed out the final details on the general capitulation of German forces in Italy, which went into effect on 2 May 1945. The withdrawal of this SS armour emboldened the Italian partisans, and when Mussolini attempted to flee to neutral Switzerland, the Resistance intercepted the Duce on 28 April 1945 near Lake Garda and promptly executed him.

THE PANZER IV

The Panzer IV became the most numerous German tank of World War II and remained in production throughout the conflict. Although the Germans did not originally intend the Panzer IV to be the main element of their armoured forces, the basic design proved so effective that it ultimately became the premier German tank of the war. German firms completed the first 35 preproduction Panzer IV tanks, the Model A, in 1936. This 17.3-tonne (17-ton) vehicle mounted a short 7.5cm KwK L/24 gun, was operated by a five-man crew and featured armour up to 20mm (0.8in) thick. A 250bhp Maybach HL 108 TR petrol engine powered the tank, which could reach a top road speed of 30kph (18mph). Its most significant design feature was that the superstructure overhung the hull sides, which enabled the Germans to up-gun the tank later in the war.

During 1939–40, the Germans manufactured 248 Model D tanks which featured heavier armour. This was the only early Panzer IV variant to serve with the Waffen-SS. Over the winter of 1939–40, Hitler ordered the expansion of the *Leibstandarte*, and in the process the unit received a single troop of Panzer IV Model D tanks. These half-dozen vehicles represented the first tanks that the Waffen-SS ever received. Subsequently, some 223 Model E tanks entered German Army service between December 1940 and March 1941. The next variant, the Model F1, was the first up-armoured vehicle that the Germans developed in response to the heavy Anglo-French tanks encountered in the West in May 1940. German firms delivered 975 Model F1 tanks between February 1941 and March 1942. The F1 tank possessed homogenous armour plate of an increased 50mm- (2in-) thickness on all frontal surfaces and 30mm (1.2in) on the sides that replaced the bolted-on applique plates retrofitted to earlier Model D and E vehicles. By the start of the invasion of the Soviet Union, the Germans fielded 548 Panzer IV Model F1 tanks.

Panzer IVs in Waffen-SS service

In March 1942, German firms commenced production of the Panzer IV Model F2 that mounted the more potent long-barrelled 7.5cm KwK 40 L/43 gun. These tanks equipped the four SS formations converted to panzergrenadier divisions during 1942. The new L/43 gun made the Model F2 a more potent vehicle than its predecessors, which soon proved itself capable of taking on the Soviet T-34 tank. The 7.5cm L/43 gun possessed a muzzle velocity of 740 m/s (2428ft/s) which allowed it to penetrate 89mm (3.5in) of sloped armour at 1000m (1094yds). The longer gun mounted on the Model F2, however, raised the vehicle's weight to 23.6 tonnes (23.2 tons) and reduced its top speed to 40kph (25mph).

During the summer of 1942 the first Panzer IV Model G vehicles entered Waffen-SS service.

Over the next six months the newly created panzer units of the *Leibstandarte, Das Reich* and *Totenkopf* Divisions received significant numbers of Model G tanks. These vehicles featured enhanced armour protection and an improved muzzle brake on the main armament. Germans factories incorporated many refinements into the Model G vehicle during its production run, including a longer 7.5cm L/48 gun from late 1942. The Germans constructed 1724 Panzer IV Model F and G tanks during 1942.

The Models H and J

In March 1943 the new Panzer IV Model H entered service. This variant incorporated further improved protection with 80mm (3.2in) of hull nose armour, which increased its weight to 25 tonnes (24.6 tons) and reduced its top road speed to 38kph (21mph). In all, German firms produced approximately 4000 Panzer IV Model H tanks between the spring of 1943 and the summer of 1944, and this variant dominated the Panzer IV-equipped tank battalions of Waffen-SS armoured divisions during 1943–44. The final Panzer IV variant, the Model J, entered service in March 1944. This tank incorporated larger fuel tanks that extended its operational range to an impressive 322km (200 miles) by road and 210km (131 miles) cross-country. The Model J also featured a modified suspension that permitted its crew to fit wider tracks (Ostkette) for winter combat. German firms continued to construct Panzer IV Model J tanks until April 1945, by which time some 2392 had been delivered. These tanks spearheaded Waffen-SS panzer IV tank battalions during the last year of the war.

In total, German factories produced 8472 Panzer IV tanks between 1936 and 1945. Throughout the war the Panzer IV was able to match its opponents, except for a brief period in late 1941 on the Eastern Front before the Germans up-gunned and up-armoured the vehicle. By mid-1944, however, even the Panzer IV Model

J tank proved inferior to the latest Allied tanks. Its main armament had inferior penetration to that of the British Sherman Firefly and Soviet JS-II tanks. Furthermore, the poorly sloped armour of the Panzer IV also rendered it vulnerable to Allied fire, and it remained a slow vehicle for its size. Despite this, the Panzer IV remained the mainstay of Waffen-SS armour for most of 1944, until surpassed by the better-armed and protected Panther in the last months of the war.

The first Panzer IV tanks to serve with the Waffen-SS – indeed the first tanks of any type to do so – were the handful of Model D vehicles allocated to the *Leibstandarte* Motorised Regiment during the winter of 1939–40. These tanks first saw action during May 1940, where they provided valuable mobile fire support as the

Leibstandarte advanced, first through Holland, then west beyond Arras towards Dunkirk, and finally south deep into the interior of France until the French Government surrendered on 25 June 1940. This armour also participated in the battles fought by the *Leibstandarte* in the Balkans and Soviet Union during 1941.

It would not be until 1942, though, that the Waffen-SS received large numbers of tanks. The élan displayed by the *Leibstandarte*, *Das Reich*, *Totenkopf* and *Wiking* Divisions in combat on the Eastern Front led Hitler to reward the Waffen-SS by ordering that these formations be

BELOW: *A column of* Das Reich *Panzer IV Model H tanks on the advance in Russia in February 1943. Note the Schürzen plates have been removed.*

upgraded to panzergrenadier divisions. During the spring and summer of 1942, therefore, these divisions each received a panzer battalion equipped with new Panzer III Model J and L vehicles and long-barrelled Panzer IV Model F2 and G tanks. In early 1943, the Germans redeployed the recently formed I SS Panzer Corps, with the *Leibstandarte*, *Das Reich* and *Totenkopf* Divisions under command, from the West to the Eastern Front. In the bitter battles fought by I SS Corps around Kharkov during the spring of 1943, the long-barrelled Panzer IV out-performed the latest Panzer III vehicles and demonstrated the qualities that quickly transformed the tank into the mainstay of the German armoured force.

PANZERS AT KURSK

In July 1943, the Panzer IV Model F2, G and H tanks of the SS Panzer Corps spearheaded the German 'Citadel' offensive at Kursk. In this attack the Germans concentrated unprecedented armoured strength on the northern and southern shoulders of a great Soviet salient that projected westward deep into the German lines. The German goal was to encircle one million Soviet troops and thus regain the strategic initiative in the East. During June 1943, the SS Panzer Corps received large numbers of Panzer IV Model H tanks as reinforcements, and at the start of the offensive on 4 July its three SS divisions fielded some 422 tanks and assault guns, of which 170 were Panzer IV tanks. These tanks formed the outer edges of the famous Panzerkeil, or armoured wedge, that centred around a core of SS Tiger tanks.

Many of these Waffen-SS Panzer IV tanks participated in the climax of the Kursk Offensive, the great clash of armour that occurred at Prokhorovka on 12 July 1943. In this, the greatest tank battle of the war, 800 Soviet tanks engaged 650 German panzers. For eight hours the battle raged, with tanks often engaging one another amid blinding clouds of dust at ranges of under 50m (55yds). Although tactically the battle was inconclusive, the German failure to effect a breakthrough shifted the initiative to the Soviets. For the Red Army had carefully concealed from the Germans the mass of reserve armour it had concentrated east of the Kursk salient. Within days the Soviets committed these reserves in a mighty counter-stroke at Orel that signalled the demise of Hitler's last chance to alter the course of the war on the Eastern Front.

HOLDING THE LINE

In the aftermath of the German débâcle at Kursk, the long-barrelled Panzer IV Models F2, G and H – plus up-gunned and up-armoured F1 tanks – replaced the Panzer III as the mainstay of Germany's armoured forces. By then the High Command had withdrawn from frontline service virtually all the older, short-barrelled Panzer IV battle tanks for use in garrison and internal security duties. In late August 1943, for instance, only the SS Panzergrenadier Division *Totenkopf* retained a solitary Panzer IV equipped with the short 7.5cm L/24 gun.

Shortly after the defeat at Kursk, Hitler redeployed the *Leibstandarte* – now under command of the redesignated II SS Panzer Corps – to Italy in response to both the Allied invasion of Sicily and the signs of wavering commitment to the Axis emanating from the Italian government. The other two divisions of II SS Panzer Corps remained in the East, conducting desperate defensive actions in the south along the River Mius. Before leaving, the *Leibstandarte* handed over its remaining 39 Panzer IV tanks to the *Das Reich* and *Totenkopf* Divisions, and collected 60 brand new Panzer IV Model H tanks in Italy upon its arrival.

Throughout the rest of 1943 and into 1944, Waffen-SS Panzer IV tanks continued to play prominent roles in the desperate German attempts to halt continuing Soviet offensives across the front. The Panzer IV tanks of the 5th

SS Panzer Division *Wiking*, for example, distinguished themselves during the bitter fighting experienced in the Cherkassy Pocket. During January and February 1944, the Soviets encircled 75,000 German troops at Cherkassy – including the 5th SS Division. On 7 February 1944, the *Wiking*'s few remaining SS Panzer IV tanks spearheaded the desperate German break-out attempt in terrible winter conditions against Soviet forces that enjoyed overwhelming numerical

Panzer IV Model J

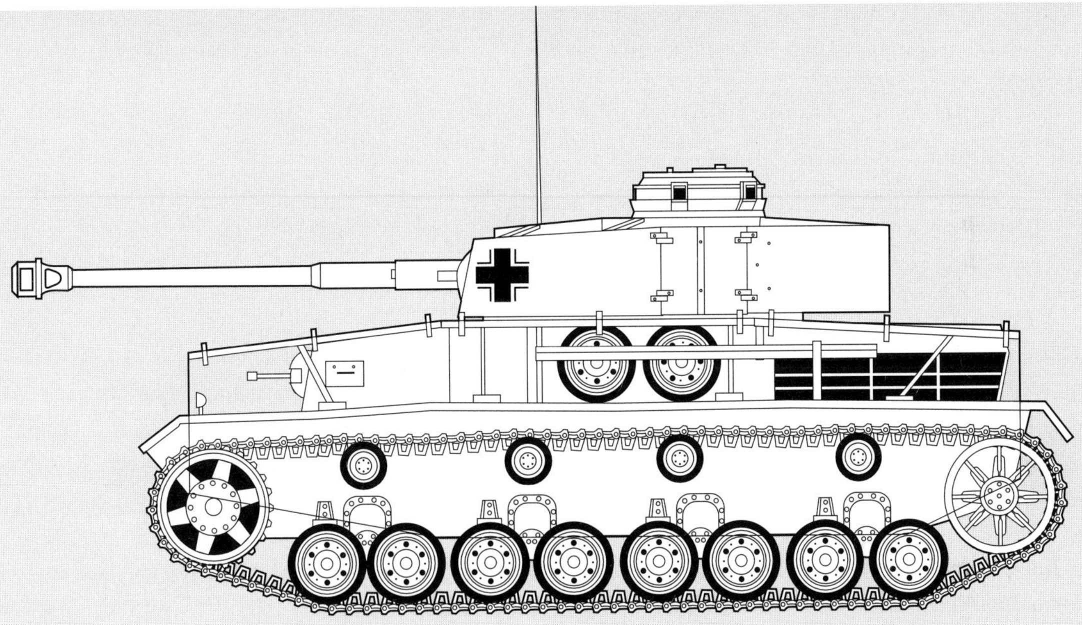

German Designation: Panzer IV Model J (Sdkfz 162/2)

Vehicle Type: Medium Tank

Crew: Four

Weight: 25 tonnes (24.6 tons)

Overall Length: 7.02m (23ft)

Hull Length: 5.90m (19ft 4in)

Width: 2.88m (9ft 5in)

Height: 2.68m (8ft 11in)

Main Armament: 7.5cm (2.95in) KwK 40 L/48 gun

Secondary Armament: 2 x 7.92mm (0.312in) MG 34

Ammunition Stowage: Main = 87 rounds; Secondary = 3150 rounds

Armour:
-Hull Front: 80mm (3.2in) (at 80 degrees)
-Hull Sides: 30mm (1.2in) (at 90 degrees)
-Hull Rear: 20mm (0.8in) (at 78 degrees)
-Turret Front: 50mm (2in) (at 79 degrees)
-Turret Sides: 30mm (1.2in) (at 64 degrees)
-Turret Rear: 30mm (1.2in) (at 74 degrees)
-Turret Roof: 20mm (0.8in) (at 0-6 degrees)

Engine: 300 bhp Maybach HL 120 TRM petrol V12-cylinder petrol

Fuel Capacity: 470 litres (103 gallons)

Maximum Speed (Road): 38kph (21mph)

Maximum Speed (Cross-Country): 16kph (10mph)

Operational Range (Road): 210km (131 miles)

Operational Range (Cross-Country): 130km (81 miles)

ABOVE: *The crew of a Waffen-SS Panzer IV Model J tank of the* Hohenstaufen *Division poses for the camera, while two locals look on from a safe distance.*

superiority. In this attempt, SS-Untersturmführer (2nd Lieutenant) Kurt Schumacher won the Knight's Cross for his outstanding bravery. Schumacher commanded two Panzer IV tanks that counterattacked a Soviet armoured company and destroyed eight T-34 tanks in the process. The next day his tank single-handedly engaged another enemy tank company. In all, during the two days fighting against overwhelming Soviet odds, Schumacher accounted for no fewer than 21 Soviet tanks!

Through super-human efforts – including forming human chains while under Soviet fire to get across the icy torrents of a river in flood spate – the exhausted troops of the *Wiking* Division

managed to break out of the Cherkassy Pocket. Almost immediately they found themselves threatened by renewed Soviet encirclement at Kovel. By the time that German reinforcements extricated the *Wiking* from Kovel, the division had become a shadow of its former self: it could muster only 3000 men and had lost all its armour. Hitler immediately ordered the division withdrawn for reconstitution and, unlike any other panzer division during the war, the *Wiking* received a two-battalion armoured regiment equipped solely with Panther tanks.

During 1944, the Panzer IV also distinguished itself during dogged defensive fighting on the Western Front as well. In France and Belgium during the spring of 1944, both the *Leibstandarte* – undergoing reconstitution after heavy losses in the East – and the new 12th SS Panzer Division *Hitlerjugend* prepared to repel the imminent

Allied invasion of Nazi-occupied France. By the start of the Allied Normandy landings on 6 June 1944, the *Hitlerjugend* fielded a full battalion of 98 Panzer IV Model H and J tanks.

Despite its increasing vulnerability to Allied fire, the Panzer IV performed well in defensive operations against the Allies in Normandy. One of the tank's finest achievements was the repulse of the Canadian advance on Carpiquet on 7 June 1944. As elements of the 3rd Canadian Infantry Division advanced south beyond the villages of Authie and Buron to extend their bridgehead, the 2nd Battalion, 12th SS Panzer Regiment, counter-attacked with 50 Panzer IV tanks and hit the exposed left flank of the Canadians in a vicious, well-coordinated counter-stroke. The Germans flung the Canadians out of Buron and Authie, inflicting heavy losses on them, though at a heavy price. At the cost of 17 Panzer IV tanks disabled or damaged, the Waffen-SS troopers counted 28 destroyed Canadian tanks abandoned on the battlefield. Flushed with victory, the fanatical Nazi youngsters callously rounded off their successful strike by executing several dozen unarmed Canadian prisoners.

PANZER IV SS ACES
The *Hitlerjugend*'s Panzer IV tanks achieved an even more resounding success just four days later. On the afternoon of 11 June, the 6th Canadian Armoured Regiment supported by the infantry of the Queen's Own Rifles of Canada advanced southwest from Norey-en-Bessin to capture Le Mesnil-Patry and the high ground beyond. The 8th Company, 12th SS Panzer Regiment, commanded by SS-Obersturmführer Hans Siegel, launched an immediate counter-attack. At the cost of just two Panzer IV tanks lost, the SS panzers annihilated the leading Canadian armoured squadron, destroying 37 Shermans. In disarray, the Canadians recoiled to lick their wounds.

One of the *Hitlerjugend* Division's most prolific tank aces in Normandy was SS-Unterscharführer (Senior Corporal) Willy Kretzschmar who commanded a Panzer IV tank of the 5th Company, 12th SS Panzer Regiment. Kretzschmar accounted for 15 Allied tanks during the desperate defensive battles that the *Hitlerjugend* conducted in Normandy. He particularly distinguished himself in heavy defensive fighting around the strategically vital Hill 112 southwest of Caen. Even though the young, ideologically motivated troops of the *Hitlerjugend* fought fanatically in Normandy, the local tactical successes achieved by their panzers could not prevent the German front collapsing in August 1944 in the face of the overwhelming combat power of the Allies. In the process of escaping from the hell of the Falaise Pocket and in the frantic retreat across France, the *Hitlerjugend* lost virtually all of its remaining tanks. On 4 September the division fielded just five operational Panzer IV tanks.

FIGHTING TILL THE END
The Panzer IV continued to play a significant role in German operations in the West throughout 1944, despite its growing battlefield inferiority. In fact, when Hitler ordered his battered forces in the west to launch a counter-stroke against the over-extended Allied positions in the Ardennes during December 1944, it was to the Panzer IV that the German field commanders first turned. The spearhead force of the Ardennes counter-offensive, SS Kampfgruppe *Peiper*, fielded a mixed force of Panzer IV and Panther tanks. For the lighter, more mobile, and less fuel-hungry Panzer IV was better suited to the rugged terrain over which Peiper's forces had to operate than the Panther and Tiger tanks available to him. In particular, the impressive operational range of the Panzer IV Model J was a vital asset in this operation because, logistically, the Germans conducted the offensive on a shoe-string, and fuel-hungry monsters like the King Tiger remained of limited value given Germany's dire fuel shortage.

During the last 18 months of the war it was not only the latest Panzer IV tanks that bore the brunt of the Waffen-SS attempts to stem the continued Allied advance on all fronts. For as the situation at the front degenerated, the Germans threw into the fray every old or obsolete tank that could be rounded up. One such training unit thrown into frontline combat in this fashion during the late summer of 1944 was the SS Panzer Training Regiment *Seelager* based at Dondangen, Latvia, which fielded five early-model Panzer IV tanks. When the Soviet 1944 summer offensive shattered Army Group Centre and ripped a huge hole in the German line, Red Army spearheads raced virtually unopposed toward the Baltic coast around Riga in an effort to cut off the German Army Group North located in Estonia. In desperation, the Germans mobilised every available unit and flung them into combat to stem the Soviet onslaught. Thus the SS Panzer Training Regiment *Seelager* found itself mobilised for combat as the SS Panzer Brigade *Gross*, named after its experienced commander, SS-Obersturmbannführer (Lieutenant-Colonel) Martin Gross, a veteran tank ace who had won a name for himself as a tank-killer at Kursk during the summer of 1943.

SCRAPING THE BOTTOM OF THE BARREL

By 8 August 1944 SS Panzer Brigade *Gross* had absorbed a variety of SS training units in the area and fielded 2500 troops and a motley array of 23 aged Panzer III and IV tanks, reinforced by seven Tiger I tanks. That day the Germans committed the Brigade to combat against an entire Soviet mechanised cavalry corps south of Libau, Latvia, with the mission of preventing a widening of the gap between the German Army Group Centre and the Axis forces in Estonia. On 20 August, SS Brigade *Gross*, now reinforced with five additional old Panzer IV training tanks, participated in a counterattack that successfully reopened an admittedly tenuous land link with the isolated German Army Group North, and which

destroyed 48 Soviet tanks in the process. The Brigade continued to fight in the Baltic states until November 1944, when it disbanded to provide desperately needed reinforcements for the four SS panzer divisions – the *Leibstandarte*, *Das Reich*, *Hohenstaufen* and *Hitlerjugend* – replenishing for the Ardennes counteroffensive.

The Panzer IV continued to play a key role in Waffen-SS defensive operations until the end of the war. The 12th SS Panzer Division *Hitlerjugend*, for example, lost 32 of its remaining 38 Panzer IV tanks in the futile German 'Spring Awakening' Offensive in Hungary during February and March 1945. Its six remaining Panzer IV tanks retreated west through Austria during April 1945 to surrender to the Americans near Amstettin in Austria in early May 1945. Likewise, the last Panzer IV Model J tanks of the 11th SS Panzergrenadier Division *Nordland* succumbed in the futile counterattack launched by SS Army Detachment *Steiner*, commanded by SS-Obergruppenführer (General) Felix Steiner, to relieve beleaguered Berlin during late April 1945. Waffen-SS tank battalions equipped with the Panzer IV thus continued to fight tenaciously until the last days of the war.

PANZER IV VARIANTS

The Germans also produced a number of specialised variants of the Panzer IV tank, the most important of which were the series of anti-aircraft artillery tanks, termed Flakpanzers. The aerial superiority that the Allies achieved on all fronts during 1943 made it imperative that the Germans develop effective anti-aircraft tanks to replace their existing improvised flak halftracks. In September 1943, Hitler authorised production of a purpose-built vehicle, the Flakpanzer IV Möbelwagen (Furniture Van). This 24-tonne (23.6-ton) anti-aircraft tank mounted a 3.7cm Flak 43 gun in a light armoured shield on top of the standard Panzer IV chassis. German firms built a total of 240 Möbelwagens between March 1944 and April 1945.

This vehicle served throughout the last year of the war in the self-propelled flak platoons of Waffen-SS panzer regiments. The 1944-pattern SS panzer division theoretically possessed eight Flakpanzer IV Möbelwagen anti-aircraft tanks in the headquarters company of its panzer regiment. Initial production of the Möbelwagen remained sluggish, however, and Waffen-SS units divisions in the field often did not receive these vehicles on schedule. In Normandy during June 1944, for example, both the *Leibstandarte* and *Hitlerjugend* Divisions fielded the stop-gap Flakpanzer 38(t) anti-aircraft tank and the latter formation only received the Möbelwagen during July 1944.

The wartime Waffen-SS also utilised a range of captured enemy tanks that it pressed into service to help offset equipment shortages. Second-rate Waffen-SS internal security and training units often deployed foreign armour and even

the élite SS panzer divisions occasionally employed small numbers of the more effective captured enemy tanks, such as the Soviet T-34. At the Battle of Kursk, for example, I SS Panzer Corps deployed 18 captured T-34s.

By 1942–43, the Waffen-SS experience with both the first generation of German tanks, as well as captured armour, demonstrated that a second generation of combat tanks was needed to fight effectively on the modern battlefield, especially to combat the hordes of Russian tanks on the Eastern Front. It is to this new wartime generation of armour with which the Waffen-SS achieved its greatest battlefield triumphs – the Panther, the Tiger and the King Tiger – that this work now turns.

BELOW: *The German Army and Waffen-SS made use of captured foreign tanks, such as these French Hotchkiss H-39 tanks being paraded in Paris.*

CHAPTER 6

Panther and Tiger Tanks

With Panther, Tiger and King Tiger tanks, Waffen-SS tank crews dominated the battlefields of the latter half of the war. But though they destroyed thousands of enemy tanks, they were never available in sufficient numbers to alter the course of the war.

The armoured fighting vehicles (AFVs) with which the Waffen-SS is most associated in the public mind are the medium and heavy tanks that Germany developed during the last half of World War II. This armour consisted of the Panzer V Panther medium tank, as well as the Panzer VI Tiger I and Panzer VIB King Tiger heavy tanks.

The German High Command only commenced serious development of a new medium tank in the autumn of 1941, after the shock of their first defeats at the hands of the formidable new Soviet T-34 medium and KV-1 heavy tanks during Operation 'Barbarossa'. The T-34 out-

LEFT: A Panzer V Panther Model G tank passes through the narrow roads of the hilly Ardennes during the German counteroffensive, December 1944.

matched any German tank then in service through the balance it struck between the requirements of protection, firepower, mobility and reliability. On 25 November 1941, the Panther commission's investigation of the nature of armoured warfare on the Eastern Front concluded that Germany urgently needed a new medium tank that incorporated the three design features where the T-34 out-matched existing German tanks: a long-barrelled, overhanging, large-calibre gun; sloped all-round armour; and a suspension with large road wheels and wide tracks for speed and mobility.

Consequently, in January 1942 Germany began to develop a 30-tonne (29.5-ton) prototype medium tank with these design features, named the Panther tank after the commission that inspired it. The prototype Panther featured

ABOVE: *The crew of a Panzer V Panther Model A tank conversing with the crew of a Sdkfz 251/6 halftracked armoured personnel carrier.*

sloped armour, a powerful 650bhp Maybach HL 210 engine, interleaved wheels with torsion bar suspension, and a hydraulically powered turret set well back to mount the new long-barrelled 7.5cm KwK L/70 gun. The Germans commenced construction of pre-production Panther tanks in May 1942, and during the summer they made massive efforts to rush the tank into large-scale production. This excessive haste resulted in a finished production vehicle that weighed 43 tonnes (42.3 tons) – well above its target weight. The earliest Panther tanks experienced numerous mechanical problems due to the vehicle's excessive weight, most notably an over-strained

gearbox and transmission. Although the Germans tried strenuously to remedy these problems, they were only able to ameliorate, rather than eliminate, these weaknesses.

The Germans commenced mass production of the Panther Model D (Sdkfz 171) in November 1942. This version featured a larger 700bhp Maybach HL 230 engine and a more resilient AK 7-200 gearbox, designed to minimise the problems associated with excessive weight that had surfaced during field trials of pre-production vehicles. Despite these teething problems, the German High Command considered the Panther critical for their future war effort, and consequently set an ambitious production target of 250 vehicles per month. The Panther Model D tank was operated by a crew of five and possessed 80-110mm- (3.2-4.3in-) thick frontal armour with 40-

45mm- (1.6-1.8in-) thick plate on the sides. From February 1944, German factories retro-fitted surviving Model D Panthers with welded 5mm (0.2in) skirt plates (Schürzen). In its first field trials, however, the Model D – despite its modifications – still experienced serious teething difficulties with over-strained gears and suspension. Hitler, however, remained determined to use the new tank as soon as possible, despite its obvious weaknesses.

Consequently, the High Command allocated all of the first 250 Panther Model D tanks produced to the two German Army panzer battalions – the 51st and 52nd – tasked to spearhead the 4 July 1943 German offensive at Kursk, codenamed Operation 'Citadel'. The three SS panzer divisions committed at Kursk – *Leibstandarte*, *Das Reich* and *Totenkopf* – therefore deployed not one Panther tank between them. Indeed, the only Waffen-SS formation to receive Model D Panthers in this period was the new 9th SS Panzergrenadier Division *Hohenstaufen*. During the first half of 1943, the Germans began raising two new Waffen-SS panzergrenadier divisions: the 9th and the 10th SS *Frundsberg*. By March 1943, the *Hohenstaufen* had received 21 Model D Panthers, although Hitler soon re-allocated these tanks to the East as last-minute reinforcements for the 'Citadel' Offensive.

POOR START AT KURSK

The Model D Panther fired its first shot in anger at Kursk on 4 July 1943, but its operational début augured ill for the future. In one Panther battalion, less than half of its 125 Panthers remained operational at the end of the first day; and by the end of the second day only 50 out of the original 250 Panthers committed were still fit for combat. Several dozen Panthers broke down after just a few hours combat due to mechanical failure – usually wrecked gears or suspensions. Engine fires caused by insufficient engine ventilation also disabled several dozen additional Panther

tanks, for the Germans had sealed the Panther's engine compartment to make it watertight to allow amphibious wading, and this often led to the engine overheating.

WAFFEN-SS PANTHERS

During late July 1943, in the aftermath of the débâcle at Kursk, the High Command terminated production of the Model D Panther because they considered it unsuitable for frontline combat. Despite this decision, German factories only completed construction of the last of the existing stocks of partly completed Model D tanks in September 1943, by which time 600 vehicles had been delivered. These late-production Model D tanks incorporated a new hull machine gun in a simple letter-box mount, as well as further modification to their gearboxes and suspension. In late July and August 1943, the SS panzer divisions *Leibstandarte*, *Totenkopf* and *Wiking* each received a single company of these modified Model D tanks. In the same period, the 1st Battalion, SS Panzer Regiment *Das Reich*, then located at training grounds in Germany, also received 49 improved Model D Panthers to equip two of its tank companies, plus its regimental staff company. In late August, the battalion rejoined the *Das Reich* Division on the Eastern Front, and promptly demonstrated the battlefield superiority of its new Panthers when they destroyed no less than 53 Soviet tanks in their first day of combat.

The initial German set-back with the Panther Model D at Kursk prompted the High Command to introduce an improved Panther Model A vehicle into service in August 1943. This tank incorporated a ball-mounted hull machine gun, a further modified gearbox and transmission, strengthened road wheels and extra coolant tubes to reduce the risk of engine fires. In total, German firms produced 1768 Panther Model A tanks between August 1943 and June 1944, at an average rate of 161 vehicles per month. During the autumn of 1943, as large numbers of Model A Panthers began to enter German service, Hitler decided that over the next year each Waffen-SS panzer division would received an entire battalion of Panther tanks to replace their current Panzer III-equipped battalion. This policy applied not just to the four existing Waffen-SS panzer divisions – 1st *Leibstandarte*, 2nd *Das Reich*, 3rd *Totenkopf* and 5th *Wiking* – but also to the 12th *Hitlerjugend* (then being raised), as well as to the forming 9th and 10th Panzergrenadier Divisions *Hohenstaufen* and *Frundsberg*, which were earmarked for conversion to full panzer divisions. In theory, a Waffen-SS Panther battalion fielded 76 Panther V tanks: four companies each with 17 tanks and a headquarters company with a further eight Panthers, including three command variants. In addition, the panzer regiment's staff company possessed a further three command Panthers, bringing the theoretical divisional total to 79 Panthers, including six command tanks.

THE FIRST SS PANTHER BATTALIONS

One of the first Waffen-SS division to re-equip with an entire battalion of Panther tanks was the 2nd SS Panzer Division *Das Reich*. By December 1943, though, this battered division had lost all its original 49 Panthers during sustained defensive operations around Zhitomir in the Ukraine. Consequently, the High Command withdrew the remnants of the division and re-equipped its 1st Panzer Battalion entirely with Panther tanks. The division redeployed to France midway through this refurbishment, and by the time the division advanced to repel the Allied Normandy landings in June 1944 it fielded 62 Panzer V tanks, most of which were Model A vehicles.

During the winter of 1943–44, the demands from the frontline for Panthers left precious few vehicles available for the 9th SS Panzer Division *Hohenstaufen*, then still in the process of forming. By March 1944, the division still had not received more than a handful of Model A Panthers. Consequently, when the High Command dispatched the *Hohenstaufen* to the east to stem the Soviet advance at Tarnopol in April 1944, the division had to leave its Panther battalion behind at the Mailly-le-Camp training park near Paris awaiting the rest of its tanks! Only in mid-June 1944, after the *Hohenstaufen* had returned to the West to halt the Allied D-Day landings, did the Panther battalion rejoin its parent formation. The unit had by then received 37 Panther Model A and G tanks – still less than half its authorised strength.

STEMMING THE RED TIDE AT NARVA

During late 1943, the 11th SS Volunteer Panzergrenadier Division *Nordland* received some Panther tanks instead of StuG III assault guns, the normal equipment for this type of SS division. In early 1944, the *Nordland* formed part of SS-Gruppenführer (Lieutenant-General) Felix Steiner's III Germanic SS Panzer Corps that held the northern part of the Eastern Front on the Gulf of Finland at Narva. The frontline in the north had changed only slightly since December 1941, although the Soviets had managed to break the German encirclement of Leningrad in 1943. In early February 1944, the Soviets launched an offensive against the *Nordland*'s defensive line along the frozen Narva River. The Soviets established a bridgehead south of the positions held by the *Nordland* and began to advance north, rolling up the SS defences from the south. In desperation, the divisional commander committed his reserve of Panther Model A tanks from the 11th SS Panzer Battalion *Hermann von Salza* – named after the Grand Master of the Medieval Teutonic Knights – to stem the Soviet advance, which they achieved. Similarly, when in late March a Soviet armoured column broke through the German defences and reached the main bridge over the Narva at Ivangorod, the *Nordland* committed the Panthers of its 1st

Panther disguised as an American M-10

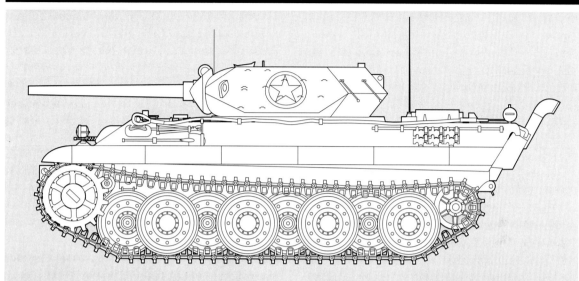

German Designation: Panzer V Panther Model G
Crew: Five
Weight: 44.8 tonnes (44 tons)
Overall Length: 8.86m (29ft)
Hull Length: 6.94m (22ft 9in)
Width: 3.27m (10ft 9xin)
Height: 3.00m (9ft 10in)
Main Armament: 7.5cm (2.95in) KwK 42 L/70 gun
Secondary Armament: 2 x 7.92mm (0.312in)
    MG 34; 1 in hull front; 1 roof-mounted
Ammunition Stowage: Main = 82 rounds;
    Secondary = 4200 rounds
Armour:
    -Hull Front (Nose): 80mm (3.2in) (at 35 degrees)

-Hull Front: 80mm (3.2in) (at 35 degrees)
-Hull Sides: 40mm (1.6in) (at 90 degrees)
-Hull Rear: 50mm (1.6in) (at 60 degrees)
-Turret Front: 100mm (4in) (at 80 degrees)
-Turret Sides: 45mm (1.8in) (at 65 degrees)
-Turret Rear: 45mm (1.8in) (at 62 degrees)
-Turret Roof: 15mm (0.6in) (at 0-6 degrees)
Engine: 700 bhp Maybach HL 230 P30 V12-cylinder
Fuel Capacity: 730 litres (161 gallons)
Maximum Speed (Road): 46kph (29mph)
Maximum Speed (Cross-Country): 30kph (19mph)
Operational Range (Road): 200km (124 miles)
Operational Range (Cross-Country): 130km
    (81 miles)

Panzer Company to frantic counterattacks against vastly superior Soviet forces; for his courage in this action SS-Oberscharführer (Senior Sergeant) Phillip Wild received the coveted Knight's Cross. III SS Panzer Corps managed to conduct an epic defence at Narva against superior odds for six months, before a massive Soviet onslaught in July 1944 forced the Waffen-SS to withdraw.

German factories commenced production of the final variant of the Panther – the Model G – in February 1944. This version incorporated features that German designers originally had planned for the new Panther II tank intended to replace the Panther. By late 1943, however, Allied aerial attacks and material shortages had forced the Germans to scale down the delayed Panther II project, and instead to incorporate some of its advanced features into the Model G Panther. The latter tank featured a redesigned hull with upper side protection thickened to 50mm (2in), and sloped instead of vertical armour on the lower-hull sides. This redesign also increased available space, which allowed main armament stowage to be increased from 79 to 82 rounds. Late-production vehicles completed during the spring of 1945 also incorporated additional features earmarked for the Panther II, such as all-steel road wheels. In total, German firms constructed 4185 Panther Model G tanks during 1944-45.

The first Waffen-SS formation to receive large numbers of Model G Panthers was the 5th SS Panzer Division *Wiking*. During February 1944, the Soviets decimated this formation in the Korsun Pocket on the Eastern Front, and the remnants withdrew to Poland for refurbishment. Here, in the spring of 1944, both the division's panzer battalions were re-equipped entirely with a total of 160 Panthers, in contravention to the policy of allocating just one Panther battalion per SS armoured division. In July 1944, the *Wiking* Division – now Germany's most potent armoured formation – was again committed to the Eastern Front in a desperate attempt to halt the extremely successful Soviet 'Bagratian' Offensive in White Russia.

PANTHERS FOR THE YOUNG SS TIGERS

Another SS formation to receive large numbers of the Panther Model G tank was the 12th SS Panzer Division *Hitlerjugend*. Formed during 1943–44, the *Hitlerjugend* was the seventh, and last, Waffen-SS armoured division raised in the war. This division consisted primarily of 17- and 18-year old Hitler Youth members – dubbed the 'candy soldiers' because of their age – together with a cadre of veteran NCOs from the *Leibstandarte* Division. The division's 12th SS Panzer Regiment comprised the standard format of one Panther-equipped and one Panzer IV-

equipped tank battalion. By D-Day, 6 June 1944, the *Hitlerjugend* deployed 81 Panthers, mainly Model G vehicles. In the bitter defensive fighting conducted by the Germans in Normandy, the division was virtually wiped out. In these battles the *Hitlerjugend*'s inexperienced but ideologically motivated young troops fought with both fanatical zeal and cruelty. In one case, for example 60 of its soldiers held out in the ruins of Falaise for three days against vastly superior Allied forces, but its young troops also callously executed at least 41 unarmed Canadian prisoners in several separate atrocities in the first weeks of the invasion.

STOPPING 'TOTALIZE'

It was in this desperate defensive fighting that the Panther tanks fielded by the *Hitlerjugend* proved their true worth. These tanks often deployed in ambush positions in woods, where the tank's mechanical failings proved less of a drawback than they would during offensive operations. Indeed, Panther tanks were prominent in one of the *Hitlerjugend*'s most impressive accomplishments of the entire Normandy campaign: its halting of the 8 August 1944 Canadian Operation 'Totalize'. After initial Allied successes on the first day, the corps commander, General Simonds, tasked Worthington Force – an all-arms battle group based around the Canadian 28th Armoured Regiment – to capture the tactically important Hill 195. As Worthington Force advanced during the night towards Hill 195, it became hopelessly lost and veered violently off course. By morning it had ended up 6.4km (four miles) northeast of its objective, and well beyond effective artillery range. Before the Canadian command realised that Worthington Force was nowhere near its objective, the Germans reacted with their customary speed and vigour. Two all-arms armoured units from the *Hitlerjugend* – SS Battle Groups *Wünsche* and *Krause* – simultaneously attacked Worthington Force from three

ABOVE: *A frontal view of a Panther Model G of the Hitlerjugend Division preparing to counterattack Anglo-American forces west of Caen, 16 June 1944.*

sides. Spearheaded by 12 Panther and 9 Tiger I tanks, the SS counterattack annihilated the isolated Canadian force in under two hours, leaving dozens of bodies and 28 burning tanks in the positions once occupied by Worthington Force.

Another Waffen-SS formation engaged in the desperate German defensive fighting conducted in Normandy was the 2nd SS Panzer Division *Das Reich*. One of the most famous Waffen-SS tank aces of the entire war – SS NCO Ernst Barkmann – served with this formation. Barkmann established his reputation with Panther tanks during late 1943 in bitter defensive combat

on the Eastern Front. In Normandy, he commanded the Model G Panthers of 4th Company, 2nd SS Panzer Regiment. The *Das Reich* Division reached the Normandy Front in late June 1944 after a protracted approach march from the south of France, during which its SS troops committed several appalling atrocities against innocent French civilians in reprisal for Resistance attacks. By far the worst excess occurred on 10

Panther Command Tank

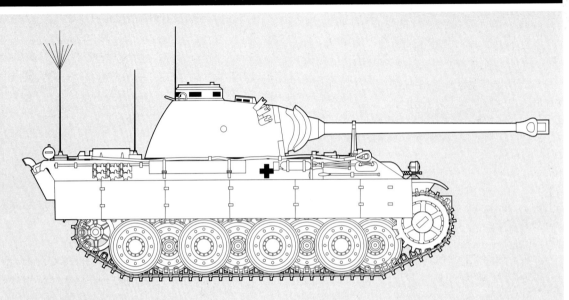

German Designation: Command Panzer V Panther
  Model G (Sdkfz 171)

Vehicle Type: Medium Command Tank

Crew: Five

Weight: 44.8 tonnes (44 tons)

Overall Length: 8.86m (29ft)

Hull Length: 6.94m (22ft 9in)

Width: 3.27m (10ft 9xin)

Height: 3.00m (9ft 10in)

Main Armament: 7.5cm (2.95in) KwK 42 L/70 gun

Secondary Armament: 2 x 7.92mm (0.312in)
  MG 34; 1 in hull front; 1 roof-mounted

Ammunition Stowage: Main = 82 rounds;
  Secondary = 4200 rounds

Armour:
  -Hull Front (Nose): 80mm (3.2in) (at 35 degrees)

-Hull Front (Driver's Plate): 80mm (3.2in)
(at 35 degrees)

-Hull Sides: 40mm (1.6in) (at 90 degrees)

-Hull Rear: 50mm (1.6in) (at 60 degrees)

-Turret Front: 100mm (4in) (at 80 degrees)

-Turret Sides: 45mm (1.8in) (at 65 degrees)

-Turret Rear: 45mm (1.8in) (at 62 degrees)

-Turret Roof: 15mm (0.6in) (at 0-6 degrees)

Engine: 700 bhp Maybach HL 230 P30 V12-cylinder
  petrol

Fuel Capacity: 730 litres (161 gallons)

Maximum Speed (Road): 46kph (29mph)

Maximum Speed (Cross-Country): 30kph (19mph)

Operational Range (Road): 200km (124 miles)

Operational Range (Cross-Country): 130km
  (81 miles)

June 1944 at Oradour-sur-Glane, where soldiers of *Das Reich* shot or burnt to death virtually the entire village's population – over 600 unarmed civilians – in four hours of unbridled savagery.

On 8 July 1944, Barkmann destroyed his first Sherman tank near the town of St. Lo, and during

the next morning he knocked out a further two American Shermans and disabled a third. That same afternoon Barkmann destroyed three more Shermans, but then had to send his Panther to the division's field workshops after it was hit by an Allied anti-tank round. Barkmann then borrowed another Panther tank to resist the continuing American offensive pressure. On 14 July, Barkmann knocked out a further three Shermans before withdrawing from the battle after an American artillery shell blew off one of his vehicle's tracks.

On 26 July, the Americans began their major breakout attempt: Operation 'Cobra'. By now Barkmann was back in his original vehicle, and took part in the desperate attempts made by *Das Reich* to stem the American onslaught. After a fierce encounter with American armour, his tank suffered a mechanical failure and only just managed to make it back to German lines. Barkmann's tank was operational again the next morning, but only after German mechanics had worked on it all through the night. The next day, the rapid American advance cut Barkmann off from the rest of his company. Despite lacking any support, he spotted and engaged an American column and single-handedly destroyed another nine Allied tanks. In the process, though, Barkmann's Panther suffered five hits which threw one of the tracks and jammed a hatch. Despite this damage, Barkmann's tank kept fighting and even managed to limp away to safety in reverse gear.

ACE OF THE KNIGHT'S CROSS

Further frantic repairs enabled Barkmann's Panther to participate again in desperate defensive combat the next day, during which the SS panzer ace notched up another six enemy kills. On 29 July, the by-now wounded Barkmann again found himself cut off from the German frontline. His tank fought its way out of yet another encirclement, but before it reached German lines Barkmann finally had to abandon his burning vehicle. Continuing their escape on foot, Barkmann's crew made their way back to the German lines. For these exploits, the German High Command awarded Barkmann the coveted Knight's Cross.

Yet even SS triumphs such as this could do little but slow the inexorable deterioration of the German frontline in Normandy. By late July 1944, growing Allied numerical superiority had worn down both the German defence and their logistical support. The sudden success of the American 'Cobra' breakout owed much to the collapse of German logistics – which also explains why the still powerful *Das Reich* failed to stem the American advance. By the second day of 'Cobra', the *Das Reich* Division was forced to abandon two entire companies of valuable Panther tanks because it had insufficient petrol for them. The success of 'Cobra' triggered the German collapse in Normandy during mid-August 1944.

PREPARATIONS FOR THE ARDENNES OFFENSIVE

After the disintegration of German resistance in Normandy, the Allies advanced rapidly into Belgium and approached the borders of the Reich itself. In the aftermath of this disaster, Hitler stunned his senior generals by ordering a counteroffensive against the weakest part of the Allied front: the rugged terrain of the Ardennes. In the month prior to the December 1944 German Ardennes counteroffensive, the High Command rebuilt the badly weakened SS panzer divisions earmarked to spearhead the offensive.

Despite these efforts, however, the four SS divisions assigned to spearhead the attack – *Leibstandarte, Das Reich, Hohenstaufen* and *Hitlerjugend* – all remained significantly understrength. Each possessed just a single organic armoured battalion with a mixture of about 90 Panzer IV and Panther tanks. The *Das Reich* Division, for example, deployed 58 Panther tanks and the *Hohenstaufen* just 35. Indeed, equipment was so scarce that the battered 10th SS Panzer Division *Frundsberg* only fielded 10 Panthers and was considered too weak to participate in the offensive.

The Waffen-SS also employed a cunning field improvisation of the Panther during the

Ardennes counteroffensive. Some 13 specially modified Panthers served with the 150th Panzer Brigade, commanded by SS-Standartenführer (Colonel) Otto Skorzeny, Hitler's favourite special operations expert. The Germans modified the Panthers to resemble the American M-10 tank destroyer by welding additional metal sheets on to their turrets and hull, and then repainted them in Allied winter camouflage schemes. The 150th Brigade was tasked with cre-

BELOW: *SS-Untersturmführer Michael Wittmann (far left) and his crew with their Tiger in Russia, 14 January 1944. Note the kill rings on the barrel.*

ating confusion and chaos behind the American frontlines, in order to aid a rapid German advance beyond the River Meuse. The formation gained modest tactical surprise initially, but subsequently, once the Allies discovered the subterfuge, the brigade not only lost all of its Panthers but failed to aid significantly the German advance to the River Meuse.

The demise of the Ardennes counteroffensive did not bring any respite for the battle-weary Waffen-SS formations of the 6th SS Panzer Army. After a few days' rest and receipt of newly produced Panthers, Hitler ordered the 6th SS Panzer Army to Hungary to participate in the 10

February 1945 'Spring Awakening' Offensive. The *Hitlerjugend* Division, for example, received 16 new Panther Model G tanks to raise its strength to 44 Panzer V tanks. After four weeks' bitter but futile fighting in Hungary against overwhelming Soviet forces, the division had lost 35 Panthers and fielded just nine operational Panzer V tanks. The Panzer V remained the principal weapon of Waffen-SS panzer units until the end of the war. It was, therefore, fitting that small groups of SS Panthers, such as those fielded by the 11th SS Volunteer Panzergrenadier Division *Nordland*, participated in the final, futile defence of Berlin during April 1945. In one of the last Panther actions of the war, the division's last two tanks spearheaded the attempt by the remnants of the encircled Berlin garrison to escape Soviet captivity on 2 May 1945. Even though both Panthers were destroyed, they helped create a tiny gap in the Soviet encirclement that allowed several hundred troops of the garrison to fight their way out west to surrender to the Americans.

SPECIALISED PANTHERS

Germans factories also constructed small numbers of several specialised Panther tank variants that served with the Waffen-SS, such as an artillery observation tank, command tanks and a recovery vehicle. From late 1943, German factories produced 41 Panzerbeobachtungs panthers: armoured artillery observation variants of the Panther tank. Prior to this, German artillery spotters had relied on armoured halftracks and specialised tanks which mounted a dummy gun instead of their main armament. This Panther variant carried sophisticated observation devices that included two periscopes and a stereoscopic range-finder mounted inside the turret. Furthermore, the Panzerbeobachtungs panther retained its main armament, and this made it more valuable in combat that earlier observation vehicles. These features made this Panther variant the most effective German

armoured observation vehicle of World War II, and only its scarcity limited the tactical impact it had on the battlefield. The Germans allocated one or two of these vehicles to select panzer divisions, where they served with the self-propelled armoured artillery battalion. The élite Waffen-SS panzer divisions received most of these vehicles. The *Hitlerjugend*, for example, fielding five of these observation Panthers – almost one-eighth of the total produced – at the start of the German counteroffensive in the Ardennes.

COMMAND TANKS

During 1943–45, German factories also out-fitted approximately 600 Panthers as Sdkfz 167 command tanks. These proved superior to earlier German command vehicles in that they retained their main armament, even though they carried less ammunition than the standard Panther tank to make space for the powerful extra radio equipment. The vehicle carried a long-distance Fu 8 radio supported by a distinctive 1.3m (4ft 3in) star antenna mounted on the rear deck, which together provided effective communication up to a distance of 65km (40 miles). Initially, only German factories could convert Panther battle tanks into command variants, but from July 1944 all new Panther tanks incorporated features that enabled divisional workshops to convert in the field standard tanks into command vehicles. Overall, the Panther command tank proved highly effective, not just because it retained its 7.5cm gun, but also because it closely resembled the Panther battle tank and was hence not easy for the enemy to identify and target. In theory, each 1944-pattern SS panzer division fielded six command Panthers, but not even the favoured SS divisions always deployed a full complement. The *Hitlerjugend*, for example, possessed just four command Panthers on 6 June 1944, and five in the Ardennes counterattack. Two other SS divisions in the Ardennes – the *Leibstandarte* and *Das Reich* – each also fielded just four command Panthers.

At Kursk, the Germans experienced serious problems recovering disabled Panthers from the battlefield due to their great weight. The artillery tractors then in German service proved inadequate for the task, and tank commanders had to use operational Panthers to recover disabled ones from the battlefield, though this often caused the towing Panther to break down as well! The Germans soon developed a purpose-built, fully tracked recovery vehicle based on the Panther chassis. Between August 1943 and April 1945, Henschel constructed 350 Bergepanther recovery vehicles. These mounted a 40-tonne (39.4-ton) winch powered by the tank's electrical turret traverse system. On top of the hull deck was mounted an open-topped steel and wood framework, together with a canvas roof, which gave limited protection to the crew from enemy fire. The Germans planned to issue two Bergepanthers to every panzer division, although Waffen-SS panzer divisions such as the *Hitlerjugend* often fielded several additional vehicles.

BIRTH OF THE TIGER

Another AFV closely associated in the popular imagination with the Waffen-SS was the Tiger tank. The famous Panzer VI Tiger I heavy tank grew out of the German projects undertaken during 1940–41 in response to the battlefield experiences gained during May 1940 in the West and during 1941 in the East. In May 1942, the High Command selected Henschel's Tiger prototype for production under the designation Panzer VI Model E Tiger I, and ordered 1500 to be constructed. Henschel commenced production of the Tiger I in August 1942, and delivered a total of 1354 Tigers up to September 1944, when construction ceased. Of this total, approximately 500 served with Waffen-SS units, first in three divi-

sional heavy tank companies, and later in three independent SS heavy tank battalions.

The massive 56-tonne (55.1-ton) Tiger I was an angular heavy tank that resembled the smaller Panzer IV. The five-man vehicle possessed 100mm- (3.9in-) frontal and 80mm- (3.2in-) side

RIGHT: *A Tiger of the 9th Heavy Tank Company, SS Panzer Regiment* Das Reich, *preparing for the Kursk Offensive in June 1943.*

and rear armour. As its main armament, the Tiger I mounted the lethal, long-barrelled 8.8cm KwK 43 L/56 gun. Early vehicles were powered by the 642bhp Maybach HL 210 engine, though later ones carried the more powerful 700bhp HL 230 engine. The tank could achieve a maximum speed of 38kph (23.5mph) by road, and 20kph (12.5mph) cross-country. Henschel also produced two specialised command tank variants of the Tiger I that were identical to the standard tank except for the addition of powerful aerials and radio transmitters.

The Tiger I first entered operational service with the German Army in August 1942 near Leningrad, but the High Command only allocated the Waffen-SS their first Tiger tanks during the winter of 1942–43. By this time the three premier SS divisions, the *Leibstandarte*, *Das Reich* and *Totenkopf*, had completed their conversion from motorised to panzergrenadier divisions, and each received a single heavy panzer company of 14 Tiger tanks to augment further their combat

Panzer VI Model E Tiger I

German Designation: Panzer VI Model E Tiger I (Sdkfz 181)

Vehicle Type: Heavy Tank

Crew: Five

Weight: 56 tonnes (55.1 tons)

Overall Length: 8.24m (27ft)

Hull length: 6.20m (20ft 4in)

Width: operational = 3.73m (12ft 3in); transport = 3.15m (10ft 4in)

Height: 2.86m (9ft 5in)

Main Armament: 8.8cm (3.5in) KwK 36 L/56 gun

Secondary Armament: 2 x 7.92mm (0.312in) MG 34; 1 co-axial in turret; 1 in hull front.

Ammunition Stowage: Main = 92 rounds; Secondary = 3920 rounds

Armour:
-Hull Front (Nose): 100mm (3.9in) (at 66 degrees)

-Hull Front (Driver's Plate): 100mm (3.9in) (at 80 degrees)

-Hull Sides: 60-80mm (2.4-3.2in) (at 90 degrees)

-Hull Rear: 82mm (3.2in) (at 82 degrees)

-Turret Front: 100-110mm (3.9-4.3in) (at 80 degrees)

-Turret Sides: 80mm (3.2in) (at 90 degrees)

-Turret Rear: 80mm (3.2in) (at 90 degrees)

-Turret Roof: 26mm (1in) (at 0 degrees- 9 degrees)

Engine: 700 bhp Maybach HL 230 P45 V12-cylinder petrol

Fuel capacity: 534 litres (118 gallons)

Maximum Road Speed: 38kph (24mph)

Maximum Cross-Country Speed: 20kph (12.5mph)

Operational Range (Road): 100km (62 miles)

Operational Range (Cross-Country): 60km (37 miles)

power. These vehicles first saw combat at Kharkov in the spring of 1943, where they demonstrated both the lethality of their 8.8cm guns and their invulnerability at normal combat ranges to most Soviet anti-tank rounds.

The mighty Tigers played a key role in the German Kursk Offensive of 4 July 1943. I SS Panzer Corps, incorporating these three élite SS divisions, fielded some 422 tanks, including 41 Tigers. Here, these armoured monsters again demonstrated their awesome killing power. The Tiger tank of SS-Unterscharführer (Senior Corporal) Hans Mennel, for example, a troop commander in the 6th Company, SS Panzer Regiment *Das Reich*, destroyed no fewer then 24 Soviet tanks in six days. Similarly, on 7 July massed Soviet armour counterattacked the German advance from the southern shoulder of the Kursk salient at Psyolknee. The solitary Tiger tank of SS-Oberscharführer (Sergeant) Franz Staudeggar of the *Leibstandarte*'s 13th Panzer Company engaged a Soviet force of 50 T-34 tanks! Staudeggar exhausted his entire ammunition supply as he desperately fought to stem the Soviet onrush. By the time that the Soviet armour withdrew, Staudeggar's Tiger had accounted for 22 Soviet tanks, a feat that earned him the Knight's Cross.

TIGERS VERSUS JOSEF STALINS

After Kursk, the *Leibstandarte* fought in Italy, while the *Das Reich* and *Totenkopf* Divisions remained heavily engaged on the Eastern Front. The Tiger company in each of these three divisions fought determinedly throughout 1943. In early 1944, the High Command withdrew these three Tiger companies from their respective divisions and sent them to training grounds in Germany. Here, they acted as the cadres for the new 101st-103rd SS Heavy Tank Battalions.

During the spring of 1944, these three SS heavy tank battalions played prominent roles in the frantic Eastern Front defensive battles fought by the Germans against numerically superior Soviet forces. The 102nd SS Heavy Tank Battalion, which formed part of II SS Panzer Corps, helped stem the Soviet onslaught around Tarnopol in April 1944. However, by mid-1944 new Soviet heavy tanks such as the Josef Stalin II began to threaten the battlefield superiority of the Tiger I. In May 1944, during the first major engagement between Tigers and Soviet JS-II tanks at Targul Frumos, Romania, the German tanks only managed to penetrate the armour of the JS-II at ranges under 1800m (1970yds), at which distance several Tigers succumbed to the powerful 12.2cm gun mounted on the Soviet tank. Consequently, losses of Tiger I tanks on the Eastern Front rose markedly.

'TIGER-PHOBIA'

In addition to their sterling efforts in the East, Waffen-SS Tiger I units also made valuable contributions to the German defence in the West during the summer of 1944. Despite the perception of Allied troops that the Waffen-SS fielded hundreds of Tiger tanks in Normandy, in reality just 90 SS Tigers fought in the campaign with the 101st SS and 102nd SS Heavy Tank Battalions. The damage inflicted on the Allies by these Waffen-SS Tigers confirmed the impressive defensive reputation of the vehicle, since in a static role the tank's mechanical unreliability was less of a hindrance. The skill and determination of SS Tiger crews helped create 'Tiger-phobia' within the Allies: a reluctance to engage SS Tigers. In one case, a solitary Tiger tank engaged Allied positions unmolested for hours because no Allied tanks dared to attack it. In an attempt to stem the spread of 'Tiger-phobia', General Montgomery banned British combat reports that recorded the prowess of SS Tigers on the grounds that these undermined morale. The British soon developed the rule of thumb that it took five Shermans to knock out a Tiger, but that only one of them was likely to return from the engagement!

The most impressive exploit achieved by Waffen-SS Tiger units in Normandy involved the greatest Tiger ace of World War II, SS-Obersturmführer (Lieutenant) Michael Wittmann. On 13 June 1944, the Tigers of Wittmann's 2nd Company, 101st SS Heavy Tank Battalion, inflicted a bloody repulse on the British 'Desert Rats' – the 7th Armoured Division – near Villers Bocage. By his death on 8 August 1944, Wittmann, now promoted SS-Hauptsturmführer (Captain), had claimed an incredible 139 kills in combat, a tally that earned him the highly coveted Knight's Cross with Swords and Oak Leaves. The encounter at Villers Bocage, the most famous Tiger engagement of the entire war, has passed down into somewhat exaggerated legend as the battle where a single Tiger tank smashed an entire British armoured brigade!

CLASH AT VILLERS BOCAGE

On 11 June 1944, the 101st Battalion reached the Normandy frontline, after redeploying from Belgium. Two days later, on 13 June 1944, Wittmann's command Tiger tank was reconnoitring near a small copse near Hill 213 when he observed British tanks of the 4th County of London Yeomanry Regiment advancing through the hazy daylight 1.6km (one mile) to the northeast of Villers Bocage. The British 22nd Armoured Brigade had just captured Villers Bocage, and was now advancing along the main road to Hill 213. Concealed in the wood, Wittmann was amazed by the complacency that characterised the British advance, as the tanks halted for a tea-break! 'They're behaving as if they've won the war already,' the vehicle's gunlayer, SS-Oberscharführer (Sergeant) Woll, snarled with contempt. 'We're going to prove them wrong,' retorted Wittmann.

Wittmann ordered his group of five Tiger tanks into action. The SS tanks moved westwards down the hill, avoiding the British armoured spearhead, and entered the eastern part of Villers Bocage, just as British vehicles of the HQ Squadron, 22nd Armoured Brigade, advanced through the town. Catching the British by surprise, Wittmann's tank quickly destroyed three Churchill tanks. Another Churchill tank bravely engaged the German vehicles, but quickly withdrew from the uneven firefight to take up an ambush position in a nearby garden. Wittmann's Tiger then turned round and advanced east back up the hill, to engage the British armoured spearhead from the rear. On his way back up the hill, Wittmann's tank spotted and destroyed the Churchill tank that had taken cover in the garden. Three of the four-man British crew managed to escape their stricken vehicle and tried to escape on foot, but as they did so Wittmann's crew callously machine-gunned them in the back, killing one of them. Emerging out of a wood, Wittmann's Tiger caught the rear of the British spearhead by surprise, and advanced firing repeatedly into the confused British armoured column. Within two minutes his tank had left 13 British vehicles burning on the road. Reinforced by Wittmann's other four Tigers, and with fire support from eight other Tigers, the German armour decimated the British spearhead. In under 15 minutes this battle had left 29 British armoured vehicles destroyed!

WITTMANN'S TIGER IS DISABLED

Wittmann's force then stopped to take on ammunition, after which he headed west back down the hill into Villers Bocage with two other Tigers. In the meantime, the British 22nd Armoured Brigade had assumed defensive positions in the town to resist Wittmann's attack. A Sherman Firefly mounting a 17-pounder gun – the only British tank that stood a chance against a Tiger – had deployed into a side street to ambush Wittmann's vehicles from the flank. Three Churchill tanks, supported by a six-pounder antitank gun, had taken up similar positions. As

Wittmann's Tiger, in the middle of his three-tank column, passed a side alley, the anti-tank gun fired at it from close range and immobilised the German vehicle. Wittmann's crew abandoned their stricken vehicle and fought their way south to reach the German lines. Next, after the lead Tiger succumbed to the fire of the Sherman Firefly, the third Tiger drove into the corner building, collapsing the masonry onto the six-pounder gun, before it retreated out of the town. Despite losing four Tigers out of the 13 involved, the Germans had inflicted 257 casualties and destroyed 47 Allied vehicles in the finest feat of armoured combat in the entire campaign in northwest Europe.

By 8 August 1944, however, the German front in Normandy was creeping closer to collapse as

ABOVE: *A column of Tigers of the 101st SS Heavy Tank Battalion advance in dispersed order towards the Normandy Front in June 1944*

the Allied forces wore down the German defences. As the Allies closed in on the town of Vire that day, the Tigers of the 1st Company, 102nd SS Heavy Tank Battalion, scored one last tactical success. The Tiger tank of SS-Unterscharführer (Senior Corporal) Willi Fey smashed a British tank column, destroying 14 of its 15 Shermans. Later that day Fey's Tiger knocked out a fifteenth Sherman with his last two rounds, though his tank was then immobilised by Allied fire and had to be towed back to a rear depot by two other Tigers. That same day, other Tigers of the 1st Company knocked out a further nine Allied tanks, for a tally of 24 Allied tanks destroyed by a single Tiger company in a single day! Temporary German tactical successes such as these, however, could not avert the catastrophe engulfing their forces in Normandy as the Allies successfully encircled them at Falaise. Very few of the 102nd SS Battalion's Tigers escaped the devastation inflicted on the Germans in the Falaise Pocket, and none managed to get back across the River Seine during the frantic German retreat of late August 1944. Yet, despite losing all its Tigers, the 102nd SS claimed the destruction of no less than 227 Allied tanks in just six weeks of combat in Normandy!

THE KING TIGER

During September 1944, the Germans terminated production of the Tiger I in favour of the superior King Tiger. By then the Tiger was both outclassed by the new Soviet JS heavy tanks, as well as vulnerable to potent new Allied anti-tank guns. The ending of production after delivery of just 1354 vehicles caused the number in service to plummet from 631 in July 1944 to just 243 by December. This decline in availability ensured that the Germans deployed just one Tiger I tank during the Ardennes counteroffensive, and this vehicle was soon destroyed – the day of the Tiger was all but over. By 30 March 1945, the Allies had inflicted such losses on the collapsing German forces that the latter now only fielded about two dozen operational Tiger I tanks. As the Allies advanced deep into the Reich during March 1945, Hitler – in desperation – ordered German training schools to march towards the gunfire. The last 10 available Tiger I training vehicles, together with their instructors, spearheaded the improvised SS Ersatz Brigade *Westfalen*, which the High Command committed at Paderborn in a

futile attempt to prevent the Allies encircling the entire German Army Group B in the Ruhr.

The last tank used operationally by the Waffen-SS during World War II was the Panzer VI Model B King Tiger. This heavy tank represented a logical development of the Tiger I that incorporated the well-sloped armour of the Panther. Always a rare vehicle, the King Tiger posed a formidable threat due to its combination of lethal

RIGHT: A King Tiger in the December 1944 Ardennes Offensive giving a ride to a squad of paratroopers, a photograph that illustrates the size of the tank.

firepower and virtual invulnerability to enemy fire. The massive 69.4-tonne (68.3-ton) tank mounted the lethal, long-barrelled 8.8cm KwK 43/3 L/71, for which it carried 84 rounds of ammunition, as well as two MG 34 machine guns for close defence. The King Tiger possessed superb protection, thanks to its 185mm- (7.3in-) thick turret-face armour, well-sloped 150mm- (5.9in-) thick hull glacis plate, and 80mm (3.2in)

side and rear armour. The tank was powered by a 700bhp Maybach HL 230 engine, which permitted a maximum speed of 38kph (24mph) by road, and 17kph (10.5mph) cross-country. Due to its voracious fuel consumption, the tank carried 860 litres (189 gallons) of fuel in its tanks, but still only achieved a maximum operational range of just 110km (68 miles) by road and 85km (53 miles) cross-country.

Henschel completed a prototype King Tiger vehicle in late 1942, and delivered the first three production models in January 1944. The March 1945 Allied capture of the Kassel factory halted Henschel's manufacture of the King Tiger, by which time it had completed 489 vehicles, including 20 command variants. These command King Tigers carried either a 30-Watt Fu 8 or 20-Watt Fu 7 transmitter in the turret, in addition to the usual 10-Watt radio, and carried less ammunition to make space for the extra communications equipment. Of these 489 King Tigers, approximately 150 fought with the Waffen-SS.

BATTLEFIELD PREDATOR

The King Tiger possessed lethal killing power: it could penetrate Allied Sherman tanks from any angle at over 3200m (3501yds). Its 150-185mm- (5.9-7.3in-) thick frontal armour also rendered it virtually impregnable to any Allied tank or anti-tank gun; indeed, there is no evidence that an Allied round ever penetrated the vehicle's frontal armour during the war! Of course, King Tigers remained vulnerable to disablement through damage to their tracks, while many were lost during German withdrawals due to lack of fuel or mechanical breakdown. The tank's enormous weight and ravenous fuel consumption rendered it a slow and immobile vehicle, a drawback exacerbated by mechanical unreliability. Thus the King Tiger performed best when used in a static fire-support role.

The King Tiger remained such a rare tank – the number in service peaked in February 1945 at just 219 tanks – that the High Command issued it to a select few independent heavy tank battalions. Just three Waffen-SS units received King Tigers: the 101st, 102nd and 103rd SS Heavy Tank Battalions, which served as corps troops attached to SS panzer corps. The first Waffen-SS unit to deploy King Tiger tanks was the 1st Company, 101st SS Heavy Tank Battalion. This company received 14 tanks in early August 1944

to replace the Tiger I tanks destroyed by the Allies in Normandy. These tanks conducted a desperate defence in the face of overwhelming Allied superiority, but all were destroyed either in Normandy or during the frantic German retreat across France and Belgium during late August and early September 1944.

DISAPPOINTMENT IN THE ARDENNES

In September 1944, the High Command redesignated these three SS units as army troops instead of corps troops and renumbered them as the 501st–503rd Battalions. Hitler ordered the 501st SS Battalion back to Germany so that it could be re-equipped entirely with new King Tigers in preparation for the proposed Ardennes counteroffensive. At the start of this attack, however, the 501st fielded just 30 King Tigers, well short of its authorised strength. The Ardennes counteroffensive is the one battle of World War II most closely linked in popular perception with the King Tiger tank. In reality, although the Germans deployed 52 King Tigers in the offensive – nearly one-third of the total then in service – their role in the attack remained modest. The Germans attached the King Tigers of the 501st SS to the *Leibstandarte* Division, and these served in the divisional spearhead, the SS Battle Group *Peiper*. The mission of SS-Obersturmbannführer (Lieutenant-Colonel) Joachim Peiper's battle group was to exploit any success as rapidly as possible before the Allies could react. The terrain over which Peiper's forces would fight was very poor, consisting of just a few narrow, winding roads in the hilly and heavily wooded terrain of the Ardennes. The ponderous mobility of the King Tiger was particularly unsuited to such a mission in this terrain. Consequently, Peiper led his battle group with a mixed force of the more mobile Panzer IV and V tanks, placing the King Tigers at the rear with instructions for them to keep up with the spearhead as best they could. They soon fell well behind.

Panzer VI Model B King Tiger

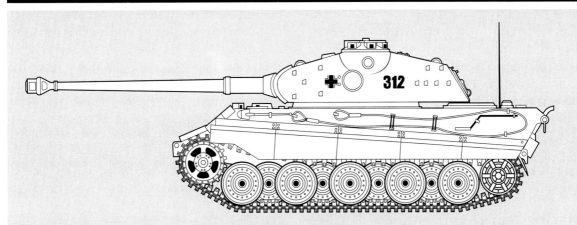

German Designation: Panzer VI Model B King
Tiger (Sdkfz 182)

Vehicle Type: Heavy Tank

Crew: Five

Weight: 69.4 tonnes (68.3 tons) first 50 Porsche
turret vehicles 69.7 tonnes (68.6 tons)]

Overall Length: 10.26m (33ft 8in)

Hull length: 7.26m (23ft 10in)

Width: operational = 3.75m (12ft 4in);
transport = 3.27m (10ft 9in)

Height: 3.09m (10ft 2in)

Main Armament: 8.8cm (3.5in) KwK 43/3 L/71 gun

Secondary Armament: 2 x 7.92mm (0.312in)
MG 34; 1 co-axial in turret; 1 hull front. Fitting for
anti-aircraft 7.92mm MG 42 on turret roof;
smoke dischargers

Ammunition Stowage: Main = 84 rounds;
Secondary: 5850 rounds

Armour:
-Hull Front (Nose): 100mm (3.9in) (at 40 degrees)
-Hull Front (Driver's Plate): 150mm (5.9in)
(at 40 degrees)
-Hull Sides: 80mm (3.2in) (at 65-90 degrees)
-Hull Rear: 80mm (3.2in) (at 60 degrees)
-Turret Front: 185mm (7.3in) (at 80 degrees)
-Turret Sides: 80mm (3.2in) (at 69 degrees)
-Turret Rear: 80mm (3.2in) (at 70 degrees)
-Turret Roof: 44mm (1.7in) (at 0-10 degrees)

Engine: 700 bhp Maybach HL 230 P30 V12-cylinder
petrol

Fuel capacity: 860 litres (189 gallons)

Maximum Road Speed: 38kph (24mph)

Maximum Cross-Country Speed: 17kph (10.5mph)

Operational Range (Road): 110km (68 miles)

Operational Range (Cross-Country): 85km
(53 miles)

Indeed, 10 King Tigers caught up with Peiper's spearhead tanks only after the battle group's advance had faltered beyond Stoumont on 20 December 1944. By the next day, however, Peiper had lost four King Tiger tanks, had also been surrounded at La Gleize by Allied counterblows and was cut off from logistic re-supply. By the night of 23-24 December, Peiper's force had run out of fuel and ammunition and was forced to fight its way out on foot, after destroying its remaining 35 tanks, including six King Tigers. The desperate nature of the fighting experienced in this part of the Ardennes attack was evidenced by the brutal massacre of 77 unarmed American prisoners by Peiper's Waffen-SS troops near Malmédy.

The collapse of Peiper's northern thrust, however, did not spell the end of the efforts demanded of the *Leibstandarte* Division by the German High Command. As the Allies began to recover, the Germans sought to maintain the initiative by switching their point of main effort farther south

BELOW: *A Totenkopf Division grenadier talking with a Hungarian soldier in Budapest, October 1944. The King Tiger is from the 503rd Heavy Tank Battalion.*

to the area around Bastogne. The town had been surrounded earlier in the offensive but determined American resistance – epitomised by the response 'Nuts!' made by the commander when invited to surrender – prevented the town falling. By Boxing Day 1944, the Allies had managed to link up with the Bastogne garrison. The German Army now redeployed the *Leibstandarte* with its remaining 17 King Tigers to Bastogne to help German efforts to re-encircle the town. In the

battles around Bastogne, the 501st SS lost another three precious King Tigers, bringing its total losses in the Ardennes to 13 King Tigers. With the inevitable defeat of Hitler's Ardennes gamble, the last great King Tiger action in the West had been fought, and lost.

EASTERN VICTORIES

Waffen-SS King Tiger tanks deployed on the Eastern Front during late 1944 not only mirrored the stunning tactical successes achieved in the West, but also proved their value in special operations designed to secure the integrity of the dissolving Axis alliance. On 20 September 1944, for example, Hitler ordered his SS commando chief, SS-Standartenführer (Colonel) Otto Skorzeny, to undertake Operation 'Panzerfaust' to ensure Hungary's continued allegiance to the Axis war against the Soviets. On 15 October 1944, the Hungarian Regent, Admiral Miklos, announced an imminent cease-fire with the Soviets, and ordered the Hungarian Army to defend Budapest from the expected Nazi backlash. For Operation 'Panzerfaust', Skorzeny's command included the attached German Army's 503rd Heavy Tank Battalion with 35 King Tiger tanks, and the élite commandos of the 600th SS Parachute Battalion. Skorzeny's forces carried out an audacious *coup d'état* to capture the Hungarian Government and thus prevent them abandoning the war against the Soviets. During the German operation the psychological impact of the sight of the massive King Tigers proved an asset to Skorzeny's troops. With the Horthy regime in Nazi custody, the Germans installed a new leader – the Fascist Ference Szalasi – who continued the Axis struggle against the Soviets (though this had almost no impact on the overall strategic situation on the Eastern Front, and only delayed the inevitable German defeat).

Such small victories were brief glimpses of sunlight in the ever-increasing gloom. The failure of the Ardennes Offensive, for example, brought no respite for the battle-weary 501st SS Heavy Tank Battalion. On 12 January 1945, Hitler ordered this unit to Hungary with the 6th SS Panzer Army to participate in Operation 'Spring Awakening'. This further, futile German offensive demanded by Hitler sought both to recapture Budapest and to protect the last, vital oil field still in Axis hands. Not surprisingly, even the determined advance of SS King Tigers soon stalled in the face of overwhelming Soviet forces. By April 1945, the inexorable Red Army advance drove the 6th SS Panzer Army back through Austria towards Germany.

THE LAST REMNANTS

Meanwhile, the 503rd SS Heavy Tank Battalion remained at training grounds in Germany awaiting new production King Tigers. By January 1945, it had received 39 of these tanks and was rushed to central Poland to help shore up the German front that had disintegrated in the face of the mighty Soviet 'Vistula-Oder' Offensive. However, nothing the German Army or Waffen-SS could do could alter the fact that the last days of the Third Reich were fast approaching. The only matter to be decided was how many would die before the German capitulation. Desperate defensive combat on the borders of the Reich reduced the 503rd SS Heavy Tank Battalion to just two operational tanks by 20 March 1945. Reconstituted one last time, the 503rd SS was finally annihilated during late April 1945 in the final, desperate German defence south of Berlin. On 22 March 1945, Hitler likewise committed his final King Tiger reserve, the 502nd SS Battalion with 31 tanks, to combat in Bohemia to bolster German attempts to stave off the final Soviet onslaught on the Reich. But during April 1945, the King Tigers of the 502nd SS – like those of its sister battalions – soon succumbed as the mighty Red Army advanced triumphantly to victory by 8 May 1945. The King Tiger had fought valiantly, but in vain, to stave off German defeat.

CHAPTER 7

SP Guns and Tank Destroyers

Assault guns and self-propelled guns proved indispensable to the Waffen-SS on the Eastern Front, but the need for greater firepower led to the to the development of purpose-built tank destroyers, which exacted a deadly toll of enemy armour.

The first armoured fighting vehicles to serve with the Waffen-SS were ironically not tanks but Sturmgeschütz assault guns. The origins of the Sturmgeschütz lay in German artillery demands for an armoured infantry support vehicle that possessed a low silhouette for better survivability, plus a dual armour-piercing and high-explosive capability. The vehicle that emerged in 1939 was armed with the short-barrelled 7.5cm KwK L/24 gun then being installed in the new Panzer IV medium tank. The requirement for a low silhouette necessitated mounting the gun in a fixed superstructure installed directly onto the chassis of the Panzer III tank. With 50mm- (2in-)

LEFT: *A propaganda photograph of the formidable Jagdpanzer V Jagdpanther which illustrates its elegant silhouette and well-sloped armour.*

thick armour on the front and 43mm (1.7in) on the sides, the vehicle possessed protection superior to that of any contemporary tank. Having designed a novel armoured fighting vehicle, the Germans designated it as an assault gun (Sturmgeschütz) to reflect its intended role to provide fire support for attacking infantry.

This vehicle, the StuG III, proved a valuable infantry support vehicle which subsequently became, when up-gunned, a potent anti-tank weapon. Designed primarily as a close support weapon for infantry, the vehicle carried a mix of high-explosive, smoke and armour-piercing rounds. However, the short 7.5cm gun could penetrate only 40mm (1.6in) of 30-degree armour at 1000m (1094yds).

During the spring of 1940, a series of pre-production vehicles underwent successful trials

StuG III Model A

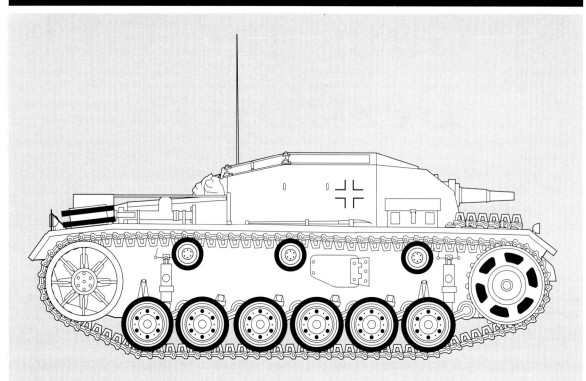

German Designation: Sturmgeschütz (StuG) III
  Model A (Sdkfz 142)
Vehicle Type: Self-Propelled Assault Gun
Crew: Four
Weight: 21.3 tonnes (21 tons)
Overall Length: 5.41m (17 ft 9in)
Overall Width: 2.92m (9ft 7in)
Height: 1.95m (6ft 5in)
Main Armament: 7.5cm (2.95in) StuK 37 L/24 gun
Main Gun Traverse: 12.5 degrees left to 12.5
  degrees right
Ammunition Stowage: 84 rounds.
Armour:
  -Hull Front (Nose): 50mm (2.0in) (at 69 degrees)
  -Hull Front (Driver's Plate): 50mm (2.0in)

(at 80 degrees)
  -Hull Sides: 30mm (1.2in) (at 90 degrees)
  -Hull Rear: 8mm (0.3in) (at 60-80 degrees)
  -Superstructure Front: 50mm (2.0in) (at 75 degrees)
  -Superstructure Sides: 30mm (1.2in) (at 90 degrees)
  -Superstructure Rear: 30mm (1.2in) (at 60 degrees)
  -Superstructure Roof: 8mm (0.3in) (at 0-12 degrees)
Engine: 300 bhp Maybach HL 120 TRM
  V12-cylinder petrol
Fuel Capacity: 320 litres (70 gallons)
Maximum road speed: 45kph (28mph)
Maximum Cross-Country speed: 19kph (12mph)
Operational range (Road): 161km (100 miles)
Operational range (Cross-Country): 97km
  (60 miles)

with five army batteries, one of which saw limited service late in the Campaign in the West. Full production commenced in July 1940, and it was during the summer of 1940 that the first SS assault gun battery came into existence. This battery received six StuG III Model A assault guns, which possessed a more powerful 320bhp Maybach HL 120 engine. But with production accorded a low priority after the stunning victories of 1940, only 184 StuG III vehicles were completed during that year, insufficient to allow the Waffen-SS to acquire more vehicles at that time.

Late in 1940 the Model B entered service, and during 1941 successive variants, the Models C, D and E, went into production with only minor modifications. German firms completed just 548 StuG III vehicles during 1941. A drawback of all these early assault guns was that they lacked a machine gun for local defence. It was during the spring of 1941 that the Waffen-SS raised additional assault gun batteries equipped with StuG III Model B-E vehicles. In the face of continuing army opposition to the SS acquiring armour, the Waffen-SS encountered less resistance obtaining 'infantry support' assault guns from the artillery. During the spring of 1941 the *Das Reich* and *Totenkopf* Divisions raised assault gun batteries, while the *Leibstandarte* absorbed the original battery. All three of these batteries participated in the German invasion of the Soviet Union.

BELOW: *An early StuG III Model A assault gun of the* Das Reich *Division on the Eastern Front, summer 1941. Note its short-barrelled 75mm gun.*

The prevalence of Soviet heavy armour on the Eastern Front quickly demanded that the StuG III be employed in an anti-tank role. Indeed, since the SS formations lacked indigenous armoured support they had to call frequently on their assault guns for offensive and defensive fire support. In light of this, on 28 September 1941, Hitler ordered that the StuG III be up-armoured and re-armed with a longer-barrelled 7.5cm gun to give it a genuine anti-tank capability.

This vehicle went into production as the StuG III Model F during the late spring of 1942, and began to reach Waffen-SS divisions during the summer of 1942. The vehicle mounted a longer 43-calibre 7.5cm StuK 40 L/43 cannon, which both necessitated modification of the frontal superstructure and increased the vehicle's over-

BELOW: *A long-barrelled StuG III Model G assault gun on the Eastern Front in September 1943. Note the five 'kills' marked on the barrel.*

all weight to 21.6 tonnes (21.3 tons). As more assault guns became available during 1942, the SS raised additional assault gun batteries, now equipped with seven StuG III Model F vehicles. During the autumn of 1942, for example, the SS Cavalry Division *Florian Geyer* received a battery. Late in 1942 there also came into existence the first SS assault gun battalion, raised for the élite *Leibstandarte* Division, equipped with 21 StuG III Model F vehicles.

In 1943 the final StuG III variant, the Model G, entered service. This mounted a more potent 7.5cm StuK 40 L/48 gun that delivered enhanced penetration of up to 91mm (3.6in) of 30-degree sloped armour, and 109mm (4.3in) of unsloped armour at 1000m (1094yds). The Model G was also the first variant to carry a 7.92mm MG 34 machine gun for local defence and it was better protected, mainly through the addition of 30mm- (1.2in-) thick applique plates bolted onto the hull front. The side and top armour was also

increased to 30mm (1.2in) and 20mm (0.8in), respectively, and side-skirts (Schürzen) were fitted to protect against hollow-charge weapons.

With production standardised on this model, assault gun production blossomed. During 1943, for example, German factories produced 3041 StuG III vehicles and a further 4973 during 1944–45. Such increased production allowed a dramatic rise in the number of assault guns in Waffen-SS service during 1943–44. This period also witnessed the introduction of an expanded

10-vehicle battery establishment and a new battalion organisation comprising 31 assault guns.

The three premier Waffen-SS divisions – *Leibstandarte*, *Das Reich* and *Totenkopf* – were reorganised in late 1942 as panzergrenadier divisions, and returned to the Eastern Front in the spring of 1943 each equipped with an assault gun battalion of 21 StuG III Model G vehicles. These participated in the counteroffensive launched by I SS Panzer Corps on 19 February 1943 that totally routed the Soviet advance in the Ukraine and

Marder II

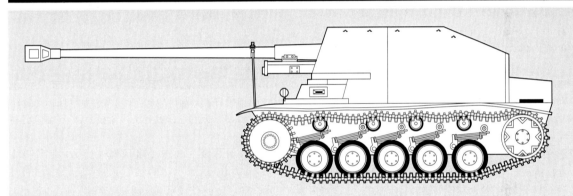

German Designation: 7.62cm Panzerjäger Marder II (Sdkfz 131)

Vehicle Type: Light Tank Destroyer (Self-Propelled Anti-tank Gun)

Crew: Four

Weight: 10.7 tonnes (10.5 tons)

Chassis: Panzer II

Overall Length: 4.88m (16 ft)

  -Hull Length: 4.64m (15ft 3in)

Width: 2.3m (7ft 6in)

Height: 2.65m (8ft 8in)

Main Armament: 7.62cm (3in) Pak 36(r) gun

Main Gun Traverse: 33 degrees left to 32 degrees right

Ammunition Stowage: 30 rounds

Armour:

  -Hull Front (Nose): 30mm (1.2in) (at 78 degrees)

-Hull Front (Driver's Plate): 30mm (1.2in) (at 90 degrees)

-Hull Sides: 15mm (0.6in) (at 90 degrees)

-Hull Rear: 8mm (0.3in) (at 90 degrees)

-Superstructure Front: 15mm (0.6in) (at 62 degrees)

-Superstructure Sides: 15mm (0.6in) (at 84 degrees)

-Superstructure Rear: none

-Superstructure Roof: none

Engine: 140 bhp Maybach HL 62 TRM R6-cylinder petrol

Fuel Capacity: 200 litres (44 gallons)

Maximum Speed (Road): 45kph (28mph)

Maximum Speed (Cross-Country): 19kph (12mph)

Operational Range (Road): 185km (115 miles)

Operational Range (Cross-Country): 121km (75 miles)

recaptured the city of Kharkov. It was at Kharkov that the rearmed StuG III first demonstrated its mettle in SS service. In a week's fighting, the SS armour and assault guns annihilated the two rifle corps and the 25th Tank Corps of the Sixth Guards Tank Army, claiming the destruction of no less than 615 Soviet tanks!

Buoyed by this success, the Waffen-SS raised additional assault gun battalions during the latter half of 1943 for the *Wiking* and *Nordland* Divisions. Over the winter of 1943–44, both the new 16th SS Panzergrenadier Division *Reichsführer-SS* and the refitted 4th SS Panzergrenadier *SS-Polizei* Division gained new assault gun battalions as well.

During 1944, however, several significant changes emerged in the way the Waffen-SS employed assault guns. These shifts reflected the changing fortunes of war as Germany was forced onto the strategic defensive. Growing shortages of tanks and delayed production of tank destroyers compelled the Germans to integrate assault gun battalions into divisional panzer regiments and panzerjäger battalions, where the StuG III served until the end of the war. During the spring of 1944, for example, it became necessary to assign two companies each of 22 StuG III Model G assault guns to the panzer regiments of the *Leibstandarte* and *Das Reich* Divisions, then refitting in France in preparation to repel the long-anticipated Allied invasion.

Another development during 1944 was the assignment of assault guns in lieu of Jagdpanzer tank-destroyers in SS divisional anti-tank battalions. The new 1944-pattern SS grenadier division was authorised a tank-destroyer company equipped with 10 Jagdpanzer 38(t) Hetzers, but production delays ensured that a number of divisions, including the 13th SS Mountain Division *Handschar*, had to make do with assault guns instead. Similarly, the 1944-pattern SS panzer division organisation envisaged an anti-tank battalion

Marder III

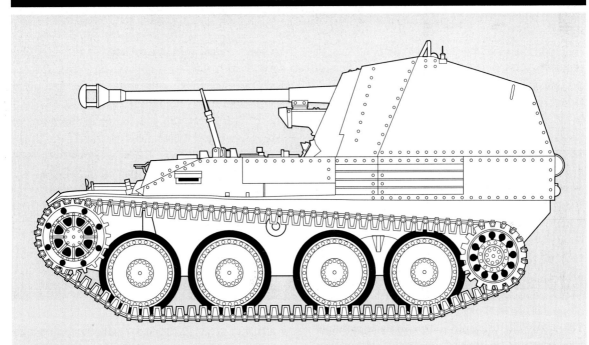

German Designation: 7.5cm Panzerjäger Marder
  III Model M (Sdkfz 138)
Vehicle Type: Light Tank Destroyer (Self-Propelled
  Anti-tank Gun)
Crew: Four
Weight: 10.5 tonnes (10.3 tons)
Chassis: Panzer 38(t)
Overall Length: 4.65m (15ft 3in)
Hull Length: 4.65m (15ft 3in)
Width: 2.16m (7ft 1in)
Height: 2.48m (7ft 10in)
Main Armament: 7.5cm (2.95in) Pak 40/3 L/46 gun
Main Gun Traverse: 30 degrees left to 30 degrees
  right
Ammunition Stowage: 38 rounds

Armour:
  -Hull Front (Nose): 20mm (0.8in) (at 80 degrees)
  -Hull Front (Driver's Plate): 25mm (1in)
  (at 78 degrees)
  -Hull Sides: 15mm (0.6in) (at 90 degrees)
  -Hull Rear: 8mm (0.3in) (at 90 degrees)
  -Superstructure Front: 25mm (1in) (at 74 degrees)
  -Superstructure Sides: 10mm (0.4in) (at 78 degrees)
  -Superstructure Rear: none
  -Superstructure Roof: none
Engine: 125 bhp Praga EPA TZJ R6-cylinder petrol
Fuel Capacity: 218 litres (48 gallons)
Maximum Speed (Road): 42kph (26mph)
Maximum Speed (Cross-Country): 24kph (15mph)
Operational Range (Road): 185km (115 miles)
Operational Range (Cross-Country): 140km
  (87 miles)

equipped with 21 Jagdpanzer IV tank destroyers. But delayed production compelled most SS panzer divisions to rely instead on assault guns until late in the year.

Mounting tank shortages also dictated the partial re-equipment of SS panzer regiments with assault guns for the December 1944 Ardennes Counteroffensive. Both the panzer regiments of the *Das Reich* and *Hohenstaufen* Divisions possessed two tank companies entirely equipped with 14 StuG III Model G assault guns. This reliance increased yet further as the German war economy crumbled during the spring of 1945. As of 15 March 1945, 99 StuG III assault guns remained operational with 13 Waffen-SS divisions, making it the most common armoured fighting vehicle (AFV) then in Waffen-SS service. By 10 April 1945 this total had declined to just 86 vehicles.

FIRE SUPPORT FOR THE INFANTRY

The increasing use of the StuG III as an anti-tank role deprived SS grenadiers of the fire support that the assault gun originally provided. Thus from 1943 the Germans diverted some 10 per cent of assault gun production to a new assault howitzer variant which mounted a short (28-calibre) 10.5cm StuH 42 L/28 howitzer. The StuH 42, as the vehicle was commonly known, lacked a machine gun for local defence and carried only 36 rounds of 10.5cm ammunition. The vehicle thus had enhanced high-explosive capability at the expense of its armour-piercing performance. German factories produced 1117 StuH 42 assault howitzers during 1943–44, and they served with select SS assault gun battalions, though in small numbers. On 15 March 1945, three SS panzergrenadier divisions – the 4th *SS-Polizei*, 23rd *Nederland* and 32nd *30 Januar* – fielded just 11 operational StuH 42 assault howitzers between them.

During the opening year of the Russo-German War, assault guns provided crucial mobile fire support to the thinly stretched and greatly outnumbered SS infantry, for whom armoured reserves were always scarce. Indeed, in this period when the SS motorised infantry divisions lacked armour, assault guns proved invaluable in an anti-tank role. In combat the StuG III proved robust, mechanically reliable and effective. Armed with the same gun as the Panzer IV and with superior protection and a lower silhouette, it could hold its own on the battlefield, although by 1945 its 7.5cm gun was barely a match for the new Soviet and Allied heavy tanks. Though the StuG III suffered from the disadvantage of lacking a rotating turret, this drawback was less of a handicap in the defensive role in which the Waffen-SS increasingly employed assault guns late in the war.

It was in these bitter defensive battles during 1943–45 that the StuG III came into its own, and revealed its true capabilities as a tank-killer. Probably the most famous SS assault gun ace was SS-Sturmbannführer (Major) Walter Kniep, who commanded the 2nd SS Assault Gun Battalion *Das Reich*, during 1943. Under his leadership, between 5 July 1943 and 17 January 1944, the battalion claimed the destruction of 129 Soviet tanks for the loss of a mere two assault guns, a record than earned Kniep the coveted Knight's Cross and his battalion a mention in dispatches.

The StuG III also provided sterling offensive service with the SS during the latter stages of the war. Assault guns figured prominently among the

SS armour committed to the Ardennes operation, where the vehicle's low silhouette and fuel economy made it an obvious choice for operations in such wooded and hilly terrain. The StuG III was thus in the spearhead of the German armour that stealthily advanced down the hilly, tortuous roads of the Ardennes, and as meagre fuel supplies dwindled it was the economical StuG III that commanders chose to keep operational.

Another example of the counteroffensive use of the StuG III was the combat of the SS Panzer

BELOW: *Dubbed 'Guderian's Duck' because of its problematic development, the Jagdpanzer IV Model F entered production during the spring of 1944.*

Brigade *Gross* on the northern sector of the Eastern Front during August 1944. This brigade was formed in just three days in early August 1944 from SS training units at the Seelager camp near Riga, and quickly committed to combat against Soviet spearheads that threatened to encircle the whole of Army Group North in Estonia and Latvia. Refitting at Riga was a company of the 1st SS Assault Gun Battalion, which was integrated into the brigade as the SS Tank-Destroyer Battalion *Gross*. On 8 August 1944, this unit's 12 StuG III assault guns successfully repulsed an entire Soviet mechanised cavalry corps south of Libau, Latvia, in bitter defensive fighting. Then, on 20 August, its assault guns recaptured Tuckums to restore landward communications with the rest of Army Group North, knocking out 48 T-34 tanks in the process. The German ability to improvise in just a few days a combined arms brigade, capable not only repulsing enemy attacks but throwing back the Red Army, reflected the flexibility and organisational skills of the Waffen-SS, and also the proven tank-killing potential of the StuG III.

FOREIGN ASSAULT GUNS

A number of foreign assault guns also found their way into Waffen-SS service. The most numerous was the Italian Semovente Da 47/32 Su Scafo L/40 light assault gun. Developed by Fascist Italy in 1942, some 250 were built. This assault gun married the chassis of the L6 light tank with a short 32-calibre 4.7cm anti-tank gun in an open box superstructure. After Italy's capitulation in September 1943, the Germans appropriated many of these vehicles. Some found their way to the 105th SS Assault Gun Battalion of V SS Mountain Corps. This unit served in the Balkans from 1943 until the spring of 1945, when the corps transferred to the Oder Front and participated in the final defence south of Berlin. By 1945 the 4.7cm gun of the L/40 was outclassed, but the vehicle's low silhouette offset this disadvantage to some degree.

During 1944 this unit received a few of the newer and better-armed Italian assault guns, including the Semovente M13/40 and M15/42, which mounted a short-barrelled 7.5cm gun. The Germans manufactured 55 of the latter at Ansaldo in northern Italy during 1944–45. As of 15 March 1945, the 105th SS Panzer Battalion deployed nine Semovente L/40 vehicles plus a solitary 75mm gun-armed Semovente M13/40. The battalion received reinforcements just prior to its redeployment to the Berlin area, and on 13 April 1945 it fielded 17 Semovente M13/40 and M15/42 assault guns. These were all lost in the Halbe Pocket south of Berlin in late April 1945.

RUSSIAN SS ASSAULT GUNS

Even smaller numbers of captured Soviet assault guns also found their way into SS hands, though the scale and extent of their use is difficult to discern because SS divisions rarely reported having captured enemy armour intact for fear that they would have to surrender such equipment to higher headquarters! However, the *Kaminski* Brigade, a force of Russian nationalists who had kept the Lokot region of Byelorussia clear of Soviet partisans, brought with it its own captured Soviet assault gun when it joined the Waffen-SS on 17 June 1944. The brigade spearheaded the brutal SS suppression of the Warsaw Uprising on 5 August 1944, but it soon lost its assault gun to a Molotov cocktail attack during the bitter street fighting that raged around Warsaw's main rail station.

Although by the outbreak of World War II Germany had created a powerful tank force capable of independent strategic operations, the offensively oriented German Army had devoted little attention to stopping enemy tanks. It was in the Western Campaign of 1940 that the German Army first realised that it needed greater anti-tank firepower in the light of its encounters with Allied heavy tanks. The German response was the first of a series of improvised light self-propelled anti-tank guns – Panzerjäger (literally

Jagdpanzer IV

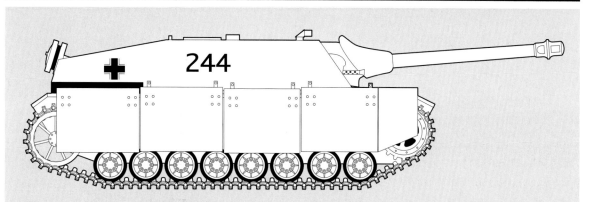

German Designation: Jagdpanzer IV
  Model F (Sdkfz 162)
Vehicle Type: Tank Destroyer (Self-Propelled
  Anti-tank Gun)
Crew: Five
Weight: 24.1 tonnes (23.7 tons)
Chassis: Panzer IV
Overall Length: 8.60m (28ft 2in)
Hull Length: 6.04m (19ft 10in)
Width: 3.28m (10 ft 5in)
Height: 1.96m (6ft 5in)
Main Armament: 7.5cm (2.95in) KwK 40 L/48 gun
Main Gun Traverse: 10 degrees left to 10 degrees
  right
Secondary Armament: 1 x 7.92mm (0.312in) MG
  34 in hull front
Ammunition Stowage: Main = 55 rounds;
  Secondary: 600 rounds

Armour:
  -Hull Front: 60mm (2.4in) (at 50 degrees)
  -Hull Front (Driver's Plate): 60mm (2.4in)
  (at 45 degrees)
  -Hull Sides: 30mm (1.2in) (at 90 degrees)
  -Hull Rear: 20mm (0.8in) (at 60 degrees)
  -Superstructure Front: 80mm (3.2in) (at 45 degrees)
  -Superstructure Sides: 40mm (1.6in) (at 60 degrees)
  -Superstructure Rear: 30mm (1.2in) (at 78 degrees)
  -Superstructure Roof: 20mm (0.8in) (at 0 degrees)
Engine: 300 bhp Maybach HL 120 TRM V12-cylin-
  der petrol
Fuel Capacity: 470 litres (103 gallons)
Maximum Speed (Road): 45kph (28mph)
Maximum Speed (Cross-Country): 24kph (15 mph)
Operational Range (Road): 210km (130 miles)
Operational Range (Cross-Country): 130km
  (80 miles)

'tank-hunters') – that mounted an anti-tank cannon behind a lightly armoured shield on a variety
of available chassis.

When Germany invaded the Soviet Union
1941, its forces soon found themselves hard-
pressed by modern Soviet armour. Indeed, the
standard German towed light and medium anti-
tank guns proved ineffective against the frontal
armour of the new Soviet T-34 medium and KV-1

heavy tanks. Moreover, as the Germans plunged
ever deeper into the Soviet Union, their forces
became thinly stretched and increasingly vulnerable to Soviet armoured counter-thrusts. All
across the front, German troops clamoured for
greater and more mobile anti-tank firepower.

In response, the Germans developed the
Marder (Marten) series of improvised light tank
hunters which carried Germany's most powerful

ABOVE: *A good view of the small size and low silhouette of the Jagdpanzer 38(t) Hetzer. It began to enter Waffen-SS service in the latter half of 1944.*

anti-tank weapons. On 22 December 1941, the Germans began development of the Marder III Panzerjäger. This vehicle mounted the captured Soviet 7.62cm M36 field gun – rechambered to take the standard German 7.5cm PzGr 40 anti-tank round – on the chassis of the Panzer 38(t) Model G tank. This superstructure comprised a small, three-sided 10mm- (0.4in-) armoured shell that afforded the gunner and loader only rudimentary protection. The 7.62cm gun was mounted well-forward and projected well beyond the front of the vehicle. This gun could penetrate 81mm (3.2in) of 30-degree sloped armour at 914m (1000yds). The vehicle also retained the secondary armament of the Panzer 38(t) tank: a 7.92mm MG 37(t) machine gun in the hull front. The Germans kept the chassis modifications to a minimum to speed conversion, and as a result the vehicle possessed a high 2.5m (8ft 2in) silhouette and weighed a substantial 10.7 tonnes (10.5 tons).

Production of the Panzerjäger 38(t) (Sdkfz 139) Marder III began on 24 March 1942, and the vehicle saw its combat début on the Eastern Front in the following month. The Marder III quickly proved its tactical worth, as it was more than capable of destroying the Soviet T-34 tank at typical combat ranges. All told German factories converted some 344 7.62cm-gunned Marder III (Sdkfz 139) vehicles during 1942.

THE 7.5CM-ARMED MARDER IIIs

The economy and success of the Marder III led to the development during the summer of 1942 of a similar vehicle – the Marder III (Sdkfz 138) – that mounted the new German 7.5cm Pak 40/3 L/46 anti-tank gun. The change of ordnance necessitated a redesign of the superstructure, with a longer and larger shield that afforded better protection for the crew. The vehicle dispensed with the lower hull and mounted a larger shield directly on the chassis to produce a lower silhouette. The vehicle also added a gun rest on the front hull to limit wear and tear on the overhanging barrel while travelling long distances. These changes increased the vehicle's weight to 11 tonnes (10.8 tons). Production began in July 1942 utilising the Panzer 38(t) Model H chassis, and German factories produced 418 7.5cm-gunned Marder III Panzerjäger during 1942–43. These served with Waffen-SS divisional anti-tank battalions on the Eastern Front and in Italy.

On both Marder III versions, however, the rear-mounted engine necessitated forward placement of the fighting compartment, which made the vehicles both front and top heavy. As a result, later-production models of the 7.5cm-gunned Marder III underwent a major redesign. A modified chassis with the engine repositioned to the centre provided more even weight redistribution. A new sloping front plate was also added that dispensed with the machine gun to allow a larger and better-protected fighting compartment, which was moved back towards the rear of the vehicle. This modified vehicle, designated the Marder III (Sdkfz 138) Model M, remained in production until May 1944, with a total of 975 being built.

In service until the end

The Marder III proved an effective improvisation that quickly married two powerful anti-tank guns with a readily available platform to provide Waffen-SS grenadiers with the mobile anti-tank capability they so desperately needed. Though vulnerable to enemy fire, the Marder III proved a valuable stop-gap that provided maximum firepower and an effective antidote to Soviet armour. While never particularly common – German factories produced only 1737 Marder III vehicles of all types – it continued to serve in frontline SS divisions until late in the war, being progressively, though never entirely, replaced during 1944 by the Jagdpanzer 38(t). Attrition inexorably reduced their numbers drastically, however, and by 15 March 1945 only two Marder III vehicles, both 7.5cm gun-armed, remained in Waffen-SS service, with the 3rd SS Panzer Division *Totenkopf*.

The Germans also turned to the dependable Panzer II chassis as a platform for another self-propelled anti-tank gun. During the spring of 1942, Alkett married the Panzer II Model D and E chassis to the same rechambered 7.62cm (3in) M36 Soviet field gun, to produce the Panzerjäger II (Sdkfz 132) Marder II. A high, box-like, 14.5mm- (0.6in-) thick armoured superstructure with steeply sloping sides replaced the original turret and superstructure. The 7.62cm gun was mounted on top of this superstructure within a three-sided gun shield of 10mm- (0.4in-) thickness, which horizontally traversed with the main armament over a range of 65 degrees.

Marders for the Waffen-SS

Despite its light armour, the Marder II still weighed 10.7 tonnes (10.5 tons) – close to the effective load limit of the Panzer II chassis powered by the 140bhp Maybach HL 62 engine. Both weight considerations and space limitations ensured that the vehicle could only carry 30 rounds for its main armament, a drawback that limited the tactical employment of the vehicle. After converting all the remaining 185 Panzer II Model D and E tanks during late spring, Alkett converted an additional 50 Model F tanks during the summer of 1942.

That same summer the *Leibstandarte*, *Das Reich*, *Totenkopf* and *Wiking* Divisions became the first Waffen-SS formations to receive the Marder II and Marder III. Both 7.62cm-gunned Marders quickly proved popular with Waffen-SS troops, and their immediate battlefield success led to additional conversions from mid-1942. These Marder II (Sdkfz 131) Panzerjäger mounted the 7.62cm Pak 36(r) within a single-section, lightly armoured, three-sided shield fitted directly onto the chassis and supported by a gun cradle, which made this vehicle's silhouette markedly different from the original Marder II. Some 531 were converted during 1942–43, and numbers served in SS divisional anti-tank battalions until late in the war.

Jagdpanzer 38(t) Hetzer

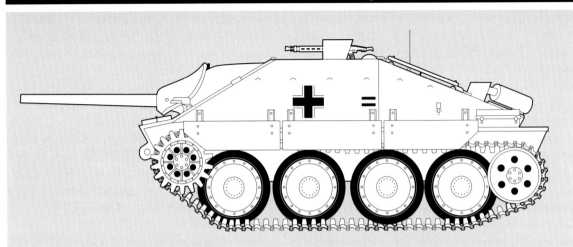

German Designation: Jagdpanzer 38(t) Hetzer (no Sdkfz #)

Vehicle Type: Light Tank Destroyer (Self-Propelled Anti-tank Gun)

Crew: Four

Weight: 16 tonnes (15.7 tons)

Chassis: Panzer 38(t)

Overall Length: 6.27m (20ft 7in)

Hull Length: 4.87m (15ft 11in)

Width: 2.63m (8ft 7in)

Height: 2.1m (6ft 10in)

Main Armament: 7.5cm (2.95in) Pak 39 L/48 gun

Main Gun Traverse: 11 degrees left to 5 degrees right

Secondary Armament: 1 x 7.92mm (0.312in) MG 34, roof-mounted

Ammunition Stowage: Main = 41 rounds; Secondary = 780 rounds

Armour:
-Hull Front (Nose): 60mm (2.4in) (at 40 degrees)
-Hull Front (Driver's Plate): 60mm (2.4in) (at 30 degrees)
-Hull Sides: 20mm (0.8in) (at 60 degrees)
-Hull Rear: 8mm (0.3in) (at 70 degrees)
-Superstructure Front: 60mm (2.4in) (at 30 degrees)
-Superstructure Sides: 20mm (0.8in) (at 75 degrees)
-Superstructure Rear: 20mm (0.8in) (at 75 degrees)
-Superstructure Roof: 8mm (0.3in) (at 0 degrees)

Engine: 158 bhp Praga EPA TZJ R6-cylinder petrol

Fuel Capacity: 386 litres (85 gallons)

Maximum Speed (Road): 26kph (16mph)

Maximum Speed (Cross-Country): 15kph (9mph)

Operational Range (Road): 161kms (100 miles)

Operational Range (Cross-Country) 80kms (50 miles)

Later conversions, using all available Panzer II chassis, carried the new German 7.5cm Pak 40/2 anti-tank gun. These vehicles, also designated Sdkfz 131, went into production in June 1942. Installation of the 7.5cm gun, however, necessitated modification of the vehicle's superstruc-

ture. The entire gun with its original shield stood on a platform positioned high on the hull. A larger, almost rectangular, protective 10mm (0.4in) shield – distinguishable from earlier variants by its slope to the rear – enclosed a larger, but still open, fighting compartment. German factories also fitted this vehicle with a barrel support mounted on the hull front to cradle the 3.5m (11ft 6in) gun during long-distance travel.

The German 7.5cm Pak 40/2 L/46 gun could penetrate 91mm (3.6in) of 30-degree sloped and 109mm (4.3in) of vertical armour at a range of 914m (1000yds), a performance superior to that of the 7.62cm Soviet gun. This Marder II variant weighed 10.8 tonnes (10.5 tons), and between 1942 and 1944 a consortium of firms converted some 1217 Marder II (Sdkfz 131) vehicles equipped with the 7.5cm Pak 40/2 gun.

The Marder II served extensively with SS divisional anti-tank battalions, particularly on the Eastern Front. During 1942–43, the Marder II and Marder III were the only effective German mobile anti-tank weapons available to counter Soviet heavy tanks, and Germany's enemies soon developed a healthy respect for the anti-tank punch of these Panzerjägers. In service the Marder II proved an effective and successful expedient. Though small and lightly protected, it packed a punch that compelled enemy armour to engage it with caution. But by 1944 a new generation of up-armed and better-protected Allied tanks had entered service, and they eroded the Marder's battlefield capabilities. Though increasingly replaced by purpose-built tank-destroyers during 1944, the Marder II continued to serve in diminishing numbers with SS formations until the very of the war.

THE MARDER I

The success of the Marder II and III led to the development of a similar vehicle, designated the Marder I. After their rapid conquest of France in May-June 1940, the Germans captured hundreds of intact AFVs. Among the most valuable of these was the Lorraine carrier: a fully tracked armoured chassis that the French had used as an infantry transporter/artillery tractor. Its powerful engine and stability impressed the Germans, who designated the vehicle as the Lorraine Schlepper.

During 1942 the Lorraine chassis appeared an obvious choice for conversion to a Panzerjäger that sported the new 7.5cm anti-tank gun. Development of the Marder I began on 25 May 1942, and during the latter half of 1942 the Germans completed 185 conversions.

The Marder I, distinguishable from its cousins – the Marder II and III – by its distinctive French wheel arrangement, carried a high, box-like, minimally armoured, open-topped superstructure that sloped to the rear. The 7.5cm Pak 40 anti-tank gun was mounted with its front shield fitted directly over this superstructure frame, which consequently limited its horizontal traverse. The long overhanging barrel also required the fitting of a barrel rest on the front hull.

A POWERFUL TANK-KILLER

The Marder I was only lightly armoured – a mere 12mm (0.5in) on the front plate and 9mm (0.35in) on the sides and rear – to keep weight down to 8.3 tonnes (7.2 tons). It was powered by a 70bhp De La Haye petrol engine that produced a maximum speed of 38kph (21mph). The vehicle held 136 litres (29.5 gallons) of fuel to give an operational range of 90-150km (56-93 miles), depending on the terrain.

The Marder I served principally with German occupation forces in France, and it was here in 1943 that the forming 12th SS Panzer Division *Hitlerjugend* acquired three of them. Likewise, the 17th SS Panzergrenadier Division *Götz von Berlichingen* received a company of 12 Marder I self-propelled guns during the spring of 1944. All of these were lost during the bitter defensive fighting in Normandy, and in the retreat across France back to the German frontier. Indeed, by mid-March 1945 there remained only six operational Marder I Panzerjäger in the entire German Army and none in Waffen-SS service.

The Marder I demonstrated its tank-killing prowess during General Montgomery's Operation 'Goodwood' offensive south of Caen,

Normandy, on 18 July 1944. Subordinated to the defending I SS Panzer Corps was Major Becker's 200th Army Assault Gun Battalion that fielded 24 Marder I vehicles. A devastating Allied aerial and artillery bombardment enabled the British 11th Armoured Division to burst through the shattered German forward defences, and dash for the commanding heights of the Bourguebus Ridge that overlooked the Falaise plain. As the 11th Armoured advanced, small groups of Marder I vehicles fought a series of delaying actions, hitting the flanks of the British spearhead then retreating, only to strike the armoured armada again. Though hugely outnumbered, these Marder I vehicles succeeded in disrupting and slowing the British advance through Cagny and Gonneville, while that evening the 1st SS Panzer Regiment assembled behind the ridge for a major counterattack that decimated the British armoured spearhead. Without loss to itself in ground combat, the 200th Assault Gun Battalion destroyed several dozen British tanks and helped to ensure that the Bourguebus Ridge remained firmly in Waffen-SS hands. So great was the damage done by the Marders that the 11th Armoured Division reported itself attacked by self-propelled 8.8cm anti-tank guns!

ENTER THE BAITER

During 1942–44, the Marder Panzerjägers fulfilled their mission of quickly providing mobile anti-tank firepower, thereby buying Germany time to develop a series of purpose-built tank-destroyers designated Jagdpanzer, literally 'hunting tanks'. The success achieved in an anti-tank role by the StuG III, with its heavy cannon mounted in a fully enclosed and well-armoured superstructure, pointed the way for future German tank-destroyer development.

During 1943 the Germans chose the dependable Czech Panzer 38(t) tank chassis as the platform for a light tank destroyer, the Jagdpanzer 38(t) Hetzer (Baiter) that was intended to replace the Marder in the anti-tank battalions of infantry divisions. The result was one of the most advanced tank destroyers of World War II. Weighing only 16 tonnes (15.7 tons), it carried a modified 7.5cm Pak 39 L/48 gun in a limited traverse mount in the front of a steeply sloped, low-silhouetted armoured superstructure on a specially widened Panzer 38(t) chassis. Its protection consisted of 60mm (2.4in) armour sloped at 40-60 degrees on the hull nose, just 8mm (0.3in) of roof and rear armour, and 20mm- (0.8in-) thick plate set at 20 degrees on the superstructure sides. Hetzer vehicles also carried 5mm- (0.2in-) thick side-skirts (Schürzen) to ward off hollow-charge rounds.

COMPACT DESIGN

The small size of the Hetzer ensured that its gun barrel significantly overhung the front of the vehicle, and the main armament was mounted off centre to the right in a distinctive Saukopf (Pig's Head) mantlet designed to minimise the possibility of deflection of incoming rounds down into the hull. The traverse of the gun was limited: only 16 degrees in both elevation and horizontal traverse. As secondary armament the Hetzer mounted a remotely controlled 7.92mm MG 34 machine gun capable of 360-degree rotation on the superstructure top, operated by the commander from inside the vehicle. The standard 150bhp Czech Praga petrol engine used in the Panzer 38(t) tank powered the Hetzer. But the heavier weight of the vehicle reduced its top speed to only 26kph (16mph) by road and 15kph (9mph) cross-country. The Hetzer had an operational range of approximately 161km (100 miles) by road and 80km (50 miles) cross-country.

The Hetzer entered production during the spring of 1944, and began to join the anti-tank battalions of SS grenadier divisions during the late summer of 1944, where it began to replace the remaining Marder Panzerjäger still in service. Once again re-equipment proved tardy, and was

never completed because the Waffen-SS initially received only 2.5 per cent of Hetzer production. The SS High Command intended to issue a tank-destroyer company of 10 Hetzers to each SS grenadier division, but it was not until September 1944 that the first SS formations began to receive the vehicle, when the assault gun battalion of the 8th SS Cavalry Division *Florian Geyer* received 29 Hetzers. During the autumn of 1944, just four SS grenadier divisions each received 14 Hetzers (the increased allotment authorised for the new 1945-pattern SS grenadier division). As Germany lurched toward defeat, fewer and fewer Hetzer's reached front-line combat units. In fact, as of 1 April 1945 the Hetzer remained on strength with only three SS divisions: the 10th SS *Frundsberg* (where it was filling in for missing Jagdpanzer IV), the 20th SS Volunteer Grenadier Division (Estonian No 1), and the 31st SS Volunteer Grenadier Division *Bohmen-Mahren*. Independent Hetzer companies were also serving with V SS Mountain Corps

and in Himmler's personal bodyguard unit, the Begleit Battalion *Reichsführer-SS*. Together these units fielded just 46 Hetzers, yet this total still made the vehicle the most numerous tank-destroyer in SS service.

The location of the Skoda and BMM works in Czechoslovakia ensured that the Hetzer was one of the last German AFVs to remain in production, and 121 were manufactured during April 1945. The last Waffen-SS formation to receive the Hetzer was the 38th SS Grenadier Division *Nibelungen*, formed in late March 1945 from the staff and pupils of the SS Officer School at Bad Tölz. It received 10 Hetzers on 15 April 1945, the last-known AFVs issued to the Waffen-SS during World War II. The *Nibelungen* Division immediately committed them to combat the following day at Neumarkt, on the Danube south of

Munich, to reinforce the shattered 17th SS Panzergrenadier Division *Götz von Berlichingen*.

An economical and efficient light tank destroyer, the Hetzer provided much-needed mobile anti-tank firepower to SS grenadiers. Its combination of fuel economy, small size and low silhouette made it well suited for the desperate defensive fighting the Waffen-SS conducted in the last year of the war. Although generally well designed, the Hetzer had a number of significant drawbacks, however, and was not a popular vehicle with SS crews. The vehicle's most serious operational limitation was the extremely restricted traverse of its main armament, the most limited of any German tank-destroyer, which meant that the entire vehicle had to be slewed to cover a target moving across the front, thus exposing the vehicle's more thinly armoured sides to the enemy. The layout of the crew compartment was also poor: the gunner and loader were positioned left of a main armament designed for right-handed operation, which reduced the Hetzer's rate of fire as the loader found it difficult to reach the ammunition supply.

HETZER OVERVIEW

Overall, therefore, though the Hetzer packed a great deal – including a heavy punch and good armoured protection – on a small chassis, it suffered from the inevitable trade-offs of such an arrangement, in terms of lack of power and speed as well as poor cross-country performance. Nevertheless, the Hetzer was a distinct improvement on the improvised Marder series of vehicles, and these improvements partially compensated for its other deficiencies. As a cheap, easily produced and fuel-efficient vehicle, the Hetzer was well-suited to Waffen-SS requirements during 1944–45.

In August 1944, Hitler ordered German firms to abandon Panzer IV tank production to concentrate instead on construction of its tank-destroyer derivative, the Jagdpanzer IV, which had entered service in January 1944. The design originated from a 1942 request for a new medium tank destroyer based on the standard Panzer IV chassis. The Army Weapons Department intended that this vehicle should carry the long-barrelled 7.5cm L/70 gun of the Panther tank in a limited-traverse mounting housed in a low-silhouetted armoured superstructure. Production difficulties, however, generated three different variants of the Jagdpanzer IV, only two of which mounted the Panther gun.

THE JAGDPANZER IV

The Army Weapons Department conceived the Jagdpanzer IV tank destroyer as an improved version of the StuG III assault gun which it would eventually replace. Heinz Guderian, the Inspector-General of Armoured Forces, however, opposed the project since he was satisfied with the StuG III, and was loath to divert current Panzer IV production capacity, then the mainstay of the German armoured force. His outspoken opposition soon earned the Jagdpanzer IV the pejorative nickname of 'Guderian's Duck', and ensured that the development of the vehicle fell behind schedule.

In addition, significant production difficulties, particularly modifying the long Panther gun to fit within the new armoured superstructure, dogged the project. Since Germany desperately needed armour at the front, an interim assault gun on the Panzer IV chassis – the Sturmgeschütz IV – was put into production. This vehicle married a modified version of the standard StuG III superstructure and its 7.5cm StuK L/48 gun directly onto the chassis of the Panzer IV tank. Between December 1943 and March 1945, German firms constructed 1139 StuG IV vehicles. Relatively small numbers of this vehicle served alongside the StuG III in SS anti-tank units during 1944–45. As of 15 March 1945, 30 operational StuG IV assault guns remained in service with five SS formations: the 5th SS Panzer Division *Wiking* and the 4th, 16th, 17th, and 23rd SS Panzergrenadier Divisions.

Jagdpanzer V Jagdpanther

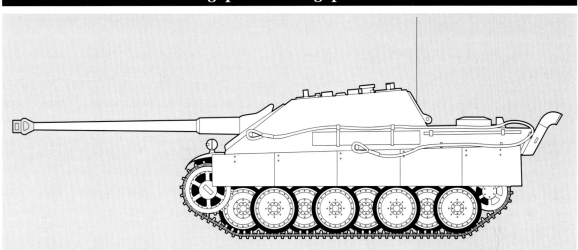

German Designation: Jagdpanzer V Jagdpanther
(Sdkfz 173)

Vehicle Type: Heavy Tank Destroyer

Crew: Five

Weight: 45.5 tonnes (44.8 tons)

Chassis: Panzer V Panther

Overall Length: 9.86mm (32ft 4in)

Hull Length: 6.87m (22ft 6in)

Width: 3.28m (10ft 9in)

Height: 2.72m (8ft 11in)

Main Armament: 8.8cm (3.5in) PaK 43/3 L/71 gun

Main Gun Traverse: 11 degrees left to 11 degrees
right

Secondary Armament: 1 x 7.92mm (0.312in) MG
34 in hull front

Ammunition Stowage: Main = 60 rounds;
Secondary: 600 rounds

Armour:
-Hull Front (Nose): 60mm (2.4in) (at 35 degrees)
-Hull Front: (Driver's Plate) 80mm (3.2in)
(at 35 degrees)
-Hull Sides: 40mm (1.6in) (at 90 degrees)
-Hull Rear: 40mm (1.6in) (at 60 degrees)
-Superstructure Front: 80mm (3.2in) (at 35 degrees)
-Superstructure Sides: 50mm (2in) (at 60 degrees)
-Superstructure Rear: 40mm (1.6in) (at 60 degrees)
-Superstructure Roof: 17mm (0.67in) (at 5 degrees)

Engine: 700 bhp Maybach HL 230 P30 V12-cylinder
petrol

Fuel Capacity: 700 litres (154 gallons)

Maximum Road Speed: 46kph (29mph)

Maximum Cross-Country Speed: 24kph (15mph)

Operational Range (Road): 210km (131 miles)

Operational Range (Cross-Country): 140km
(87 miles)

Continued delay in modifying the long 7.5cm Panther gun compelled the Germans to commence production of the Jagdpanzer IV (Sdkfz 162) in January 1944, utilising 7.5cm KwK 40 L/48 guns as fitted in the StuG III and Panzer IV. This vehicle had 60mm (2.4in) armour on the front, 30mm

(1.2in) on its superstructure sides, 5mm- (0.2in-) spaced applique plates on the rear superstructure sides, and Schürzen side skirts. In addition, it was well-sloped to maximise shot deflection.

The Jagdpanzer IV weighed 24.1 tonnes (23.7 tons) and was powered by the standard 300bhp

Maybach HL 120 engine of the Panzer IV tank. The German High Command intended to issue 21 Jagdpanzer IVs to each panzer division, but this goal proved illusory, with only 769 Jagdpanzer IV (Sdkfz 162) vehicles being completed between January and November 1944.

FIRST DEPLOYMENTS OF THE JAGDPANZER IV

The 12th SS Panzer Division *Hitlerjugend* was the first SS formation to receive the Jagdpanzer IV, during the spring of 1944. On D-Day it fielded 10 Jagdpanzer IVs, though it received further vehicles during the Normandy Campaign. During July 1944, the 3rd and 9th SS Panzer Divisions re-equipped with the Jagdpanzer IV; in September the 10th SS Panzer Division *Frundsberg* received an allotment; and the 4th *SS-Polizei* Panzer-grenadier Division and 2nd SS Panzer Division *Das Reich* followed during the autumn.

In August 1944, the Jagdpanzer IV finally received the longer 7.5cm L/70 Panther gun originally intended for the vehicle. The Germans designated this up-gunned Jagdpanzer IV as the Panzer IV/70(V), erroneously implying that it was a tank rather than a tank destroyer. The larger gun necessitated some minor changes to the vehicle, including stronger interlocking hull plates for added strength. The modified 7.5cm StuK L/70 gun carried by the Panzer IV/70(V) was ballistically identical to the 7.5cm KwK 42 L/70 Panther tank gun and overhung the hull front by 2.58m (8ft 6in), which made the 25.8-tonne (25.4-ton) vehicle nose-heavy. It carried 55 rounds of 7.5cm ammunition and also possessed a ball-mounted 7.92mm MG 34 in the hull front for local defence.

German factories produced 930 Panzer IV/70(V) vehicles between August 1944 and March 1945. It was the 1st and 12th SS Panzer Divisions in the West that first each received 21 L/70-gunned Jagdpanzer IV during October 1944 to re-equip their tank-destroyer companies. During November 1944, the 2nd and 9th SS Panzer Divisions received the vehicle, and by the start of the December 1944 Ardennes Offensive the five SS panzer divisions in the West deployed 82 Jagdpanzer IVs of both types in their divisional anti-tank battalions.

The engagements fought by the 12th SS Anti-tank Battalion in the Ardennes typified the capabilities of the Jagdpanzer IV. The Ardennes proved a taxing first combat trial for the tank destroyer, since the terrain hampered effective use of the vehicle. Indeed, in the early stages of the offensive on 17 December 1944, the battalion suffered a major blow with the death of its most prolific tank hunter, Knight's Cross holder SS-Oberscharführer (Senior Sergeant) Roy, who by that date had reached a tally of 36 tank kills in his Jagdpanzer IV since D-Day. Roy made the mistake of popping his head out of the turret of his Jagdpanzer IV to gain his bearings in the confined woods, only to be shot in the head by an American sniper. Subsequently, the battalion lost eight of its 21 Jagdpanzer IV vehicles in ceaseless, small engagements supporting SS panzergrenadiers during the counteroffensive.

FINAL VARIANT OF THE JAGDPANZER IV

Limited production ensured that it was not until the spring of 1945 that the Panzer IV/70 began to fulfill its intended role as a replacement for the Panzer IV in SS panzer battalions. Many of these vehicles were the final variant of the Jagdpanzer IV: the Panzer IV/70 Zwischenlösung (interim) or IV/70(A) which entered production in August 1944. German firms developed this special stop-gap version of the Jagdpanzer IV to overcome existing production delays. The Germans achieved this through the simple expedient of mounting the 7.5cm StuK 42 L/70 gun and a modified IV/70 superstructure on top of the Panzer IV Model J tank chassis, which was then in mass

RIGHT: *A battery of Sdkfz 124 Wespe self-propelled light field howitzers in improvised firing dugouts. The gun was mounted on the chassis of the Panzer II.*

Wespe

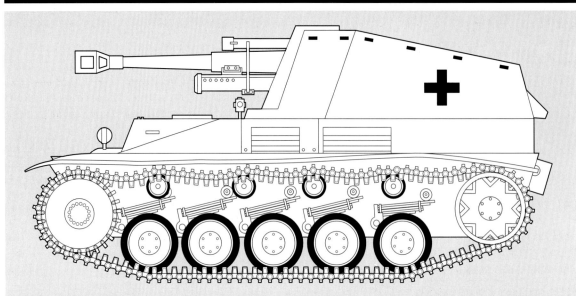

German Designation: Wespe (Sdkfz 124)

Vehicle Type: Self-Propelled Artillery Gun

Crew: Five

Weight: 11.5 tonnes (11.3 tons)

Chassis: Panzer II

Overall Length: 4.79mm (15ft 8in)

Hull Length: 4.79m (15ft 8in)

Width: 2.24m (7ft 4in)

Height: 2.32m (7ft 7in)

Main Armament: 10.5cm (4.1in) leFH 18/2 L/28 howitzer

Main Gun Traverse: 17 degrees left to 17 degrees right

Ammunition Stowage: Main = 32 rounds

Armour:
  -Hull Front (Nose): 20mm (0.8in) (at 75 degrees)

-Hull Front (Driver's Plate): 20mm (0.8in) (at 60-75 degrees)

-Hull Sides: 15mm (0.6in) (at 90 degrees)

-Hull Rear: 20mm (0.8in) (at 90 degrees)

-Superstructure Front: 12mm (0.5in) (at 69 degrees)

-Superstructure Sides: 10mm (0.4in) (at 73 degrees)

-Superstructure Rear: 10mm (0.4in) (at 74 degrees) or none

-Superstructure Roof: none

Engine: 140 bhp Maybach HL 62 TR R6-cylinder petrol

Fuel Capacity: 170 litres (37 gallons)

Maximum Road Speed: 40kph (25mph)

Maximum Cross-Country Speed: 20kph (12.5 mph)

Operational Range (Road): 140km (87 miles)

Operational Range (Cross-Country): 95km (59 miles)

production, to produce a finished vehicle more rapidly. The modified superstructure of the interim Panzer IV/70(A) possessed a cut-back rear superstructure and a vertical rear plate, making it easily distinguishable from the Panzer IV/70(V). At 28 tonnes (27.6 tons) the Panzer IV/70(A) was the heaviest of the Jagdpanzer IV. This was largely due to enhanced protection: 80mm (3.3in) on the hull nose and driver's plate, 30mm (1.2in) side and rear armour, and 120mm

(4.7in) on the gun mantlet. The vehicle had a maximum speed of 38kph (24mph) and an operational range of 322km (200 miles). The Nibelungenwerke factory in Austria produced 278 Panzer IV/70(A) vehicles between August 1944 and March 1945.

TOO FEW TO COUNT

With a low silhouette, well-sloped armour, good mobility and lethal firepower – especially the later L/70-gunned vehicles – the Jagdpanzer IV proved an efficient tank destroyer. From the spring of 1944 it progressively replaced Marder panzerjäger in the anti-tank battalions of SS panzer divisions. The Jagdpanzer IV remained an uncommon vehicle, though, and only 1977 vehicles of all three types were completed. An incomplete inventory of the German Army on 1 April 1945 recorded that 275 Panzer IV/70 of both varieties remained in operational service. The vehicle's powerful armament made it a formidable defensive weapon, particularly in the West against less well-armoured Anglo-American tanks, but like many German AFVs the Jagdpanzer IV was not produced in sufficient numbers to have a significant tactical impact. None of the three Jagdpanzer variants were particularly common in SS service. By mid-March 1945, the original Jagdpanzer IV remained in service with just four SS divisions – *Totenkopf*, *SS-Polizei*, *Wiking* and *Frundsberg* – which possessed between them 25 Jagdpanzer IV, 13 of which were operational. On that same date 108 Panzer IV/70, of which 43 were operational, were on strength with five SS divisions (the 1st, 2nd, 10th, 11th and 12th).

The last tank destroyer to see service with the Waffen-SS in very limited quantities was the superlative Jagdpanzer V Jagdpanther. In mid-1943, Germany began development of this heavy tank destroyer that mounted the powerful 8.8cm Pak 43/3 L/71 gun on the chassis of the Panther tank. The Jagdpanther comprised a standard Panther Model G tank chassis fitted with a well-sloped armoured superstructure with 80mm (3.2in) frontal, 50mm (2in) side and 40mm (1.6in) rear armour. The Jagdpanther's steeply sloped superstructure and sleek, low profile not only provided the vehicle with excellent protection but made it a particularly elegant vehicle.

The Jagdpanther weighed a substantial 45.5 tonnes (44.8 tons), but its powerful 700bhp Maybach HL 230 engine gave the vehicle good mobility, as well as an impressive 45kph (28mph) maximum road performance and a top cross-country speed of 24kph (15mph). This performance owed much to the vehicle's interleaved road wheel suspension and wide tracks, which produced ground pressure lower than that of the StuG III assault gun, a vehicle half the weight! Indeed, the balance achieved between firepower, protection and mobility made the Jagdpanther one of the best AFVs of World War II.

JAGDPANTHER – THE SUPERB BEAST

The Jagdpanther, however, remained a very rare tank destroyer and only 382 vehicles were built. From mid-1944 the it served with independent army anti-tank battalions, and in small numbers with the Waffen-SS during the last months of the war. As the 6th SS Panzer Army transferred to Hungary to mount a relief attempt to rescue the encircled garrison of Budapest during January 1945, the SS Panzer Divisions *Das Reich*, *Hohenstaufen* and *Frundsberg* each received a company of 14 Jagdpanthers. These 42 vehicles were committed in the thick of the fighting in the abortive effort to recapture Budapest and secure access to Hungary's last oil field. By 1 April, however, the number still operational had decreased to just 12. These were thrown into bitter defensive fighting in early April in a futile effort to prevent the fall of Vienna, and by 10 April 1945 just six SS Jagdpanthers remained operational.

The Germans ultimately recognised that one of the weaknesses of their panzer divisions was the lack of self-propelled (SP) artillery capable

of keeping pace with fast-moving armoured spearheads. During 1941–42, the Germans increasingly relied on the defensive power of artillery on the thinly manned Eastern Front, which further highlighted the German lack of a self-propelled howitzer. The result was the development of two self-propelled guns during 1942 that remained the backbone of German armoured artillery until the end of the war.

The first of these vehicles was the Wespe (Wasp), production of which commenced during 1942. The Wespe mounted the standard 10.5cm light field howitzer in a thinly armoured, open-topped, box-like superstructure atop the standard Panzer II chassis. The vehicle was only lightly protected, with 10mm (0.4in) armour on the superstructure and 18mm (0.7in) on the hull. Weighing 11.5 tonnes (11.4 tons), the 140bhp

LEFT: *A Hummel self-propelled gun of the* **Hohenstaufen** *Division is loaded on a flatbed railway car for strategic deployment to the East.*

Wespe self-propelled guns were built. It served in the armoured artillery battalions of SS panzer divisions. Such battalions comprised two light batteries of six Wespe vehicles and one heavy battery equipped with six Hummels. Each Wespe battery also possessed a gun-less Wespe munitions carrier that stowed an additional 90 rounds of ammunition. German firms produced some 158 of these Wespe carriers. The *Leibstandarte* became the first SS division to field the Wespe during the autumn of 1942. It received an armoured artillery battalion equipped with three batteries of four Wespe vehicles, the initial Waffen-SS organisation prior to the introduction of the Hummel. The *Das Reich* and *Totenkopf* Divisions received similarly organised battalions over the winter of 1942-43.

The Hummel (Bumble-Bee) mounted the standard 15cm heavy field howitzer in a lightly armoured rear fighting compartment atop the hybrid Panzer III/IV tank chassis. This heavy self-propelled gun weighed a substantial 25.9 tonnes (23.5tons), and to limit weight the Hummel carried just 18 15cm rounds and possessed only 20mm (0.8in) of armour on the lower hull and 10mm (0.4in) on the superstructure. Between December 1942 and late 1944, some 666 Hummel self-propelled guns were constructed. From the spring of 1943 a single battery of six Hummels plus one gun-less Hummel munitions carrier were allocated to Army and Waffen-SS panzer divisions.

The variety of AFVs that fought with the Waffen-SS not only illustrates the organisational flexibility of the armed SS, but also reflects the increasingly desperate combat in which Waffen-SS formations found themselves; situations that necessitated the commitment to combat of all available AFVs, irrespective of age, ancestry or combat effectiveness.

Maybach HL 62 engine gave it a top road speed of 40kph (25mph), but only 20kph (12mph) cross-country. The vehicle had an operational range of 200km (125 miles) by road and 113km (70 miles) over rough terrain. The large gun mounted on a small chassis also limited ammunition stowage to only 32 10.5cm rounds. The howitzer possessed a maximum range of 4120 metres (13,500yds). Between 1942 and late 1944, 683

CHAPTER 8

Artillery

Artillery – field, mountain and light, plus howitzers – provided much of the firepower that was at the core of both the offensive and defensive successes the Waffen-SS achieved during the war, though there were never enough weapons to go round. This meant SS divisions frequently had to augment their artillery units from captured stocks.

The evolution of artillery within the Waffen-SS illustrates the intense political rivalry that existed between the SS and the German Army as the latter sought to retain its right to be the sole bearer of arms within the Nazi state. When the Waffen-SS formally came into existence on 2 March 1940, its three divisions – the *SS-Verfügungs*, *Totenkopf* and *SS-Polizei* - each possessed an artillery regiment equipped with 36 light field howitzers, but they entirely lacked heavy artillery.

The first SS artillery regiment emerged in 1939 to serve with the *SS-Verfügungs* Division drawn from the SS Verfügungstruppe special

LEFT: *A light 7.5cm Gebirgsgeschütz 36 mountain gun of a mountain artillery battery deployed amid deep winter snow.*

readiness units first raised in the mid-1930s. This three-battalion motorised artillery regiment possessed 36 105cm light field howitzers, and served under army command in the September 1939 Polish Campaign, subordinated to the improvised Armoured Division *Kempf*. It participated in the German advance from East Prussia towards Warsaw, during which it provided dedicated fire support. After the campaign, post-action reports recognised the necessity for the armed SS to obtain heavy artillery to destroy fortified enemy defences and to conduct counter-battery fire.

But the army continued to resist SS efforts to obtain heavy artillery, and it took Hitler's personal intervention on 23 March 1940 to override army opposition and generate the first SS heavy artillery units. Hitler authorised heavy motorised field howitzer battalions for the *SS-Verfügungs* and *Totenkopf* Motorised Infantry Divisions, a horse-drawn heavy field howitzer battalion for the *SS-Polizei* Infantry Division (then closely identified with, but not yet integrated into, the Waffen-SS), and a motorised battalion of 10cm heavy guns for the *Leibstandarte*. However, a combination of equipment shortages and contin-

BELOW: *A 10.5cm leFH 18 light field howitzer being loaded on a landing barge by engineers of the SS-Verfügungs Division in the summer of 1940.*

ued army resistance ensured that both the *Totenkopf* and *SS-Polizei* battalions had initially to equip with captured Czech heavy howitzers.

The Rheinmetall-developed 10.5cm leichte Feldhaubitze 18 (leFH 18) light field howitzer entered service in 1935 and was the standard field piece of the light batteries. It weighed 1985kg (4377lb) and fired a 14.8kg (32.6lb) shell at a muzzle velocity of 470m/s (1542 ft/s) to a maximum range of 10,675m (11,675yds). More than 5000 were in German service in September 1939, and it remained the standard light division-

al howitzer throughout the war. In the middle of the war the modified leFH 18M, fitted with a muzzle brake and a more powerful propellant charge that increased range to 12,325m (13,480yds), entered service. In combat, however, the leFH 18 proved rather heavy, and consequently in 1941 the Germans fitted the howitzer's barrel to the carriage of the 7.5cm Pak 40 anti-tank gun to produce the leFH 18/40. This marriage of convenience only reduced the howitzer's weight by a mere 30kg (66lb), however, insufficient to improve its mobility.

A revolutionary successor, the leFH 43, entered SS service in very small numbers during the last months of the war. Though heavier than its predecessor – it weighed 2200kg (4851lb) – it had a higher muzzle velocity of 610m/s (2002 ft/s), which increased range to 15,000m (16,410yds). Instead of a standard split-trail carriage, the gun had four outriggers, two of which folded up underneath the barrel during transportation. This ingenious mechanism reduced weight while allowing a 360-degree traverse. Very few are believed to have reached SS formations in the field before the end of the war, however. The leFH 18 was the most common howitzer to serve with the Waffen-SS, and formed the core of SS divisional artillery. Waffen-SS grenadier divisions were authorised 36 leFH 18, but in the latter stages of the war it proved impossible to maintain establishment strengths. Some of the SS mountain divisions also had a single light field howitzer battalion on establishment.

LIGHT FIELD GUNS

During the later stages of the war the Germans revived use of 7.5cm light field guns after their inter-war switch to the larger 10.5cm calibre. The Germans had designed the 7.5cm leichte Feldkanone 18 (leFK 18) between the wars, and it remained in limited production throughout the 1939–45 war. The field gun weighed 1120kg (2464lb) and fired a 5.8kg (12.8lb) shell to a range of 9425m (10,318yds). It served primarily in SS cavalry formations, where its mobility proved particularly useful. Later in the war it also began to replace infantry guns in regimental cannon companies. Because it was cheap and easy to produce, the Germans stepped up production of the gun in the last year of the war, and in 1945 the weapon began to be issued to SS grenadier divisions in place of light field howitzers. The Germans also came to the realisation that a dual-purpose 7.5cm gun that could be employed in both an artillery and an anti-tank role was highly desirable. In an effort to rationalise increasingly scarce resources, the Germans introduced the 7.5cm FeldKanone 40 (FK 40) during the spring of 1945, which was a slightly modified 7.5cm Pak 40 re-equipped for dual artillery use. The SS received relatively few of these guns, however, as the German war economy collapsed under the strain of six years at war.

The 15cm schwere Feldhaubitze 18 (sFH 18) entered service in 1934, and remained the principal German heavy field howitzer throughout World War II. It was of conventional design, with a split-trail carriage and weighed 5512kg (12,154lb). It could fire eight different propellant charges depending on the range and effect desired, but the gun's effective maximum range was 13,250m (14,500yds). It was the intention of the Reichsführer-SS, Heinrich Himmler, to allocate one heavy field howitzer battalion to every SS division, but wartime realities sometimes prevented this goal from being realised. In addition, SS mountain and cavalry divisions only occasionally used the howitzer because it was really too heavy to be employed in mountainous or rough terrain. Nevertheless, the sFH 18 remained the second most common artillery piece in SS service and served until the end of the war.

HEAVY WAFFEN-SS GUNS

A rarer divisional weapon was the 10cm heavy gun, which was produced in three slightly different models: the 10cm schwere Kanone 18, 18/40 and 40. A long-calibred gun of extended range, the 10cm sK18 was intended for counter-battery and interdiction fire. As such it was produced only in limited quantities. The first SS formation to employ the gun was the élite *Leibstandarte* Regiment, which received a battalion of 12 guns in 1940. During 1942 and 1943 more of these heavy guns entered SS service: the SS Mountain Division *Nord* received a battery of four guns, as did each of the seven SS panzer divisions: *Leibstandarte, Das Reich, Totenkopf, Wiking,*

Hohenstaufen, Frundsberg and *Hitlerjugend*. In addition, later in the war a number of SS corps artillery units used the gun. Considering its rarity – the number in service peaked at 760 in June 1941 and declined thereafter – the SS received a disproportionate share of this weapon during the latter stages of the war.

ARTILLERY TACTICS

A most unusual employment of the 10cm gun was in a direct-fire role. During the German counterattack to retake the Arnhem bridge from the 'Red Devils' of the 1st British Airborne Division on 18 September 1944, the two remaining 10cm sK 40 guns of the 10th SS Panzer Artillery Regiment deployed in Arnhem park and fired over open sights at point-blank range into the buildings held by the British troops. While this was not a doctrinally approved employment of the gun – it put a valuable weapon at risk – it nevertheless proved highly effective since the gun's high muzzle velocity and relatively flat trajectory made it lethal when so employed. The defenders of Major Frost's 2nd Parachute Battalion had no answer to this weapon which demolished building after building, progressively shrinking the British bridgehead until the remaining troops had ultimately to surrender, despite offering three days of stubborn resistance.

The last 10cm heavy guns to see service in the Waffen-SS during World War II were those drawn from SS artillery training units stationed in the Protectorate of Bohemia-Moravia, the rump Czech state. In late March 1945, these units were mobilised for combat duty and dispatched to the 6th SS Panzer Army in Austria, then fighting desperately to prevent the Soviet capture of Vienna. An improvised heavy motorised artillery battalion equipped with 12 10cm sK 18 joined the SS Brigade *Trabandt*, named after its commander, SS-Obersturmbannführer (Lieutenant-Colonel) Wilhelm Trabant. This formation was one of three ad hoc brigade-sized formations mobilised in Bohemia-Moravia. Despite the firepower its heavy guns afforded it, when the brigade was committed at Zisterdorf in Slovakia it proved unable to achieve anything more than slow the Soviet steamroller. With supplies critically low and with the progressive collapse of German command, control and communications, the guns were quickly lost. On 7 May 1945, the remnants of the brigade fled westward in a desperate attempt to surrender to the Western Allies rather than the Soviets. Like all Waffen-SS forces, the brigade was anxious to avoid the revenge that they knew the Red Army would wreak in retaliation for the numerous terrible excesses that the SS had committed in the Soviet Union.

A variety of specialised mountain artillery intended for service in alpine terrain also served in the Waffen-SS. The two most common mountain artillery pieces were the 7.5cm Gebirgsgeschütz 36 (GebG 36) mountain gun and the 10.5cm Gebirgshaubitz 40 (GebH 40) mountain howitzer. These weapons served in special mountain artillery battalions. Rheinmetall developed the 7.5cm GebG 36 in the mid-1930s, and it entered service in 1938. It was of modern design with rubber-tyred wheels, split trails and a muzzle brake. Lightweight at 750kg (1654lb), the gun could be broken down into eight loads for easy transportation. It fired a 5.75kg (12.7lb) shell at a muzzle velocity of 475m/s (1559 ft/s) to a range of 9150m (10,010yds). The first SS division to receive the gun during 1942 was the newly expanded SS Mountain Division *Nord*, deployed on the Salla sector of the Eastern Front in Finland. The Division had an authorised strength of 24 7.5cm GebG 36 guns, but it was not until the summer of 1943 that the division received its full allotment of the weapon.

EQUIPPING *PRINZ EUGEN*

During the spring of 1942, the SS Volunteer Mountain Division *Prinz Eugen* then forming in the Balkans also received a battalion equipped with eight 7.5cm GebG 36. The following winter

10cm Kanone 18

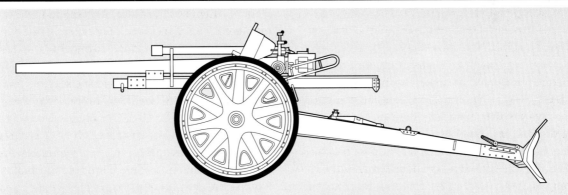

German Designation: schwere 10cm Kanone 18 (s 10cm K 18)
Weapon Type: Heavy Gun
Calibre: 10.5cm (4.1in)
Length (of piece): 5.46m (17ft 11in)
Weight: 5624kg (12,441lb) = 5.64 tonnes (5.55 tons)

Traverse: 60 degrees
Elevation: 0 to +45 degrees
Muzzle Velocity: 835m/s (2738ft/s)
Maximum Range: 19,015m (20,802yds)
Rate of Fire: 6rpm

ABOVE: *A 15cm sFH 18 heavy field howitzer accompanies German troops as they enter Danzig during the early stages of the Polish Campaign, 1939.*

the 13th SS Volunteer Croatian Mountain Division *Handschar*, raised largely from Bosnian Muslims, received two similarly organised battalions. Though it was never a common gun in Waffen-SS service, the 7.5cm GebG 36 continued to serve with SS mountain divisions until the end of the war. It proved particularly valuable in the vicious anti-partisan war the Germans conducted in the Balkans, where troops appreciated the gun's mobility and ease of transportation amid the mountainous terrain of the Yugoslavian Alps.

The 10.5cm GebH 40, produced by Böhler of Austria, became the standard German mountain howitzer of World War II. It entered service in 1942 and was of orthodox construction, with split-trail design and a weight of 1660kg (3660lb). The howitzer could be broken down into five

loads for transportation in mountainous terrain. It fired a 14.5kg (32lb) shell at a muzzle velocity of 565m/s (1854 ft/s) to a maximum range of 6740m (7378yds). The GebH 40 saw limited service in SS heavy mountain howitzer battalions. The first two SS formations to receive the gun were the SS Mountain Division *Nord* and SS Volunteer Mountain Division *Prinz Eugen*, each of which received 12 GebH 40 during 1942. In the following year the SS High Command also issued eight of the howitzers to the 13th SS Croatian Volunteer Mountain Division *Handschar*, and the weapon continued to serve alongside its smaller cousin until the end of the war.

A rarer SS mountain gun was the 7.5cm GebK 15. This modified 1918-vintage Austrian design entered production in 1935 at the Skoda Works in Pilsen, Czechoslovakia, and equipped Czech mountain formations. The gun had distinctive wheels, a large shield and a bulky barrel. For ease of transportation the gun disassembled to seven loads and when reassembled weighed 630kg (1389lb). It fired a 5.47kg (12lb) shell at a muzzle velocity of 386m/s (1267 ft/s) to a range of 6625m (7284yds). The gun was taken over by the Germans after their occupation of Czechoslovakia, but they soon deemed it unsatisfactory – one of the few confiscated Czech weapons that failed to please their German masters. The poor impression the gun made on the German Army allowed the enterprising SS to get its hands on a few dozen of them! During the summer of 1942, for example, the SS Mountain Division *Nord* received a battalion of 10 7.5cm GebK 15 mountain guns as substitutes for unavailable 7.5cm GebG 36 pieces. Very few, though, remained in SS hands during the later stages of the war.

Recoilless guns

Another specialised type of weapon that was issued principally to a select few SS cavalry formations was the compact and easy-to-transport light, recoilless gun. One of the fundamental problems of field artillery was that much of the weight of a artillery piece came from the sizeable recoil mechanism. To create a light, mobile and versatile artillery piece, it was necessary to produce a weapon without a recoil mechanism; in other words, a recoilless gun. German experiments between the wars produced a design concept that the army camouflaged under the innocuous title of 'nozzle gun'. This was a recoilless gun that had a large venturi, or nozzle, mounted at the rear, through which the gases produced on firing were exhausted, obviating the need for a recoil mechanism. In the late 1930s, after the need for secrecy was past, the Germans designated this class of artillery as light guns.

The first light gun to see operational service was the 7.5cm Leichtgeschütz 40 (LG 40), which entered service in 1941. It carried a short 18-calibre barrel and was built in two slightly different forms. The Rheinmetall version came on a carriage with motorcycle wheels, the Krupp on aircraft tail wheels. The wheels on both models were removed before firing, and the gun placed instead on a portable tripod. Weighing a mere 145kg (320lb), the LG 40 fired a 5.8kg (12.8lb) shell at a muzzle velocity of 350m/s (1149 ft/s) to a range of 6800m (7439yds). The drawback of the LG 40 was the large rearward sheet of flame it produced on firing, which proved dangerous for unwary personnel standing behind the gun! The back-blast also stirred up large clouds of dust and dirt, which frequently gave away the gun's location to an observant enemy. An additional drawback of recoilless guns was that they used five times at much propellant as conventional artillery. It was this fact that brought a decline in the production and employment of recoilless artillery in the latter stages of the war as German resources became ever more thinly stretched.

THE LG 40 IN WAFFEN-SS USE

The LG 40 served principally with Luftwaffe paratroop formations and in a small number of independent army artillery units. The gun also served with the 500th and 600th SS Parachute Battalions, each of which had a light gun platoon equipped with two LG 40s. These were intended to be air-landed by gliders, but they were very rarely used in that role. Indeed, probably the only occasion that the gun was used in an offensive airborne mission by the SS was during the 1943 Operation 'Knight's Move' designed to neutralise Josef Tito, the head of the Yugoslav Resistance, at his mountain cave headquarters outside Drvar. In this operation, apparently at least one LG 40 was glider-landed in support of the raid and provided valuable fire support for the badly outnumbered SS paratroopers.

Krupp and Rheinmetall also developed larger 10.5cm light guns, designated as the 10.5cm LG 42 and 10.5cm LG 40 respectively. The two guns were slightly different in design. The LG 42 weighed 540kg (1191lb) and mounted a 17.5-calibre barrel; the LG 40 weighed 388kg (856lb) and possessed a shorter 13-calibre barrel. Though the former generated a higher muzzle velocity, the performance of both guns were similar: each fired a 14.8kg (32.6lb) bomb to a range of 7950m (8697yds). The usual towing vehicle for all light guns was the NSU Kettenkrad halftrack motorcycle.

The 8th SS Cavalry Division *Florian Geyer* received a battalion of 12 10.5cm light guns during 1943. These light, mobile pieces were well-suited for the rapid pursuit, reconnaissance and anti-partisan role intended for SS cavalry. Rather than the normal Kettenkrad, however, horse teams pulled these guns. Their high propellant consumption and powerful back-blast ensured, however, that these weapons were not a great success, and later SS cavalry divisions received more conventional, and just as mobile, 7.5cm field guns instead.

SS CORPS ARTILLERY

In addition to divisional artillery, the SS raised a variety of corps- and army-level artillery units during the latter half of the war. As the Waffen-SS expanded during 1943, it activated its first corps commands to control the proliferating number of SS formations. This development led to the emergence of the first corps-level heavy artillery units. These were numbered in the 100 series, taking the corps number plus 100. Thus the 101st SS Heavy Artillery Battalion, for example, belonged to I SS Panzer Corps. Himmler hoped to raise a heavy artillery battalion for each SS corps, but this proved an illusory goal. In fact, as Waffen-SS resources became ever more stretched during 1944, the concept of permanently attached corps troops lost favour, and in September 1944 Himmler removed most of the SS corps troops from their respective corps headquarters and renumbered them as independent army troops in the 500 series. Thus the 101st SS Heavy Artillery Battalion, for example, became the 501st SS.

It was at corps and army level that the heaviest SS artillery was found. In such units served the 17cm Kanone 18 in Mörserlafette, a long-range, counter-battery gun mounted in a 'mortar-style' carriage that allowed high elevation. Indeed, the carriage was of advanced design that included a sophisticated dual-recoil system, in which, in addition to the barrel recoil, the whole platform also recoiled along its carriage rails. This allowed a single man to quickly rotate the gun through 360 degrees. The gun entered limited production at Hanomag of Hanover during 1941, but it was never a common weapon. Two batteries of four 17cm guns served in the 101st SS Heavy Artillery Battalion. It proved to be the best German heavy gun of the war, and could fire a 68kg (149.6lb) high-explosive round to an impressive range of 28,000m (30,652yds).

The largest calibre artillery piece to see action with the Waffen-SS was the 21cm Mörser 18 (Mrs 18). Only a few dozen served in select SS heavy artillery batteries at corps and army level. It utilised the same carriage as the 17cm K 18, but in service its range proved significantly less than that of its smaller sibling and as a consequence production ceased in 1942. Nonetheless, the gun continued in SS service until the end of the war.

SS CORPS ARTILLERY IN ACTION

In September 1944, Himmler redesignated the five existing heavy corps artillery battalions as army artillery held directly under the authority of the Reichsführer-SS. The 501st SS Heavy Artillery Battalion was replenished in western Germany during the autumn of 1944 after suffering heavy losses in the Normandy fighting, and re-equipped with two batteries of 17cm K 18 and a battery of 21cm Mrs 18. These participated in the December 1944 Ardennes Offensive, providing long-range counter-battery and harassing fire. Thereafter, the battalion redeployed with the 6th SS Panzer Army to Hungary and supported the German 'Spring Awakening' Offensive. The

21cm Mörser 18

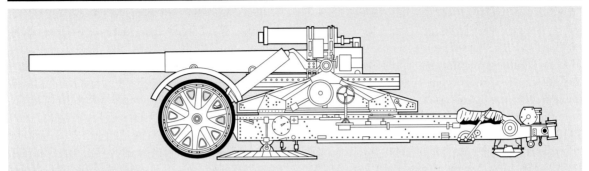

German Designation: 21cm Mörser 18 (21cm Mrs 18)
Weapon Type: Heavy Gun
Calibre: 21.1cm (8.3in)
Length (of piece): 6.51m (21ft 4in) [= L/31]
Weight: 16,700kg (36,740lb) = 16.7 tonnes
 (16.4 tons)

Traverse: 360 degrees (on platform); 16 degrees
 (on carriage)
Elevation: 0 to +70 degrees
Muzzle Velocity: 565m/s (1852ft/s)
Maximum Range: 28,000m (30,362yds)
Rate of Fire: 1rpm

502nd SS Heavy Artillery Battalion was similarly chewed up in the retreat across France during August 1944, and underwent replenishment near Prague that October before returning to the West to participate in the Ardennes Offensive. It received two batteries of six heavy field howitzers and a battery of three 21cm Mrs 18. The battalion also transferred to Hungary during February 1945, and participated in the German offensive there. The 503rd SS Heavy Artillery Battalion, which was organised like the 502nd, provided dedicated fire support for III (Germanic) SS Panzer Corps on the northern sector of the Eastern Front. The 504th SS Heavy Artillery Battalion served with IV SS Panzer Corps in Hungary. It fielded two batteries of four 15cm sFH 18 howitzers and a battery of three 21cm Mrs 18. The 505th SS Heavy Artillery Battalion fought with the V SS Volunteer Mountain Corps, first in the Balkans and then on the Oder Front during the last weeks of the war. It comprised two batteries of four 15cm sFH18 and a battery of four 10cm K 18 guns.

The Waffen-SS also raised one army level regiment: the 509th SS Mountain Artillery Regiment. This was raised from cadres of the 13th SS Volunteer Mountain Division *Handschar*, which disbanded during the autumn of 1944. The regiment possessed two battalions of 12 7.5cm GebG 36 mountain guns, a battalion of 12 10.5cm GebH 40 mountain howitzers, and a motorised battalion with eight sFH 18 and four 10cm guns. The regiment fought in Hungary with IX SS Mountain Corps and was destroyed in the Budapest Pocket during January 1945.

UTILISING CAPTURED ARTILLERY

A combination of army resistance to SS expansion and production shortfalls dictated that the Waffen-SS frequently had to utilise captured enemy artillery, belying its popular image as a well-equipped élite. An important source of arms for the SS, especially early in the war, was the rich Czech arsenal that fell into German hands as a result of the Nazi conquest of the rump Czech state in March 1939. This allowed the SS access to a range of quality Czech ordnance. During the spring of 1940, the newly activated *Totenkopf* Division received 36 10cm Czech light field howitzers, designated in German service as the 10cm leFH 4/19(t). This howitzer was a quality weapon of very similar capabilities and performance to the standard German light field howitzer. The division retained these guns until more leFH 18 howitzers became available. In the spring of 1940, the heavy artillery battalion of the *Totenkopf* Artillery Regiment also received 12 15cm sFH 25(t) Czech heavy field howitzers. Another Skoda product, this gun entered service in 1925 and saw widespread German service in the early years of World War II. The division kept these guns until June 1941, when they were replaced by standard German sFH 18 howitzers just prior to Operation 'Barbarossa'.

CZECH MOUNTAIN GUNS

As valuable were Czech mountain guns, one of the specialties of the Skoda works at Pilsen. The SS pressed into service the Czech 10cm horska houfnice vz 16, which the Germans designated as the 10cm Gebirgsgeschütz 16(t). It weighed 1235kg (2717lb), had a range of 9280m (10,159yds) and a rate of fire of between six and eight rounds per minute. It was a rather large and heavy weapon for mountain warfare, which could only be broken down into three rather hefty loads that were carried in two-wheeled carts drawn by horses or mules. Despite these drawbacks, the howitzer was used in small numbers by the Waffen-SS. During the autumn of 1943, for example, the forming 13th SS volunteer Mountain Division *Bosnia-Herzogovina* received two of these mountain howitzers. Though the SS acquired a number of these guns direct from Austria and Czechoslovakia, these two particular pieces were appropriated from Italian mountain units disarmed by the Germans in the Balkans in the aftermath of the September 1943 Italian capitulation. The Italians had in turn received the

guns as reparations booty from Austria-Hungary in 1919!

The June 1941 invasion of the Soviet Union saw the SS acquire several Red Army artillery pieces. Among these were the 12.2cm schwere Kanone 390/2(r). This was a robust and effective heavy gun that was captured in significant numbers. While relatively uncommon in the Waffen-SS, it did serve in a number of SS divisions and also at GHQ level. The 30th SS Volunteer Grenadier Division (Russian No 2), for example, raised during the late summer of 1944 from Russian auxiliaries of the Order Police, was supposed to receive 24 of these 12.2cm guns. However, the division never reached its authorised strength, and only fielded eight 12.2cm guns by the time it transferred to the Belfort sector of the Western Front in late August 1944. The division had a short and problematic history. Unwisely committed to anti-partisan operations

against the French Resistance, who were interrupting the German retreat from southwestern France, the division was rocked both by several mutinies and large-scale troop desertions into neutral Switzerland. Consequently the division was disbanded and its artillery battalion transferred to the SS White Russian Brigade, formed from the division's more reliable Byelorussians (who bitterly resented serving in a 'Russian' formation). The forming brigade was in turn redesignated as the 30th SS Grenadier Division (White Russian No 1) on 9 March 1945, but never completed formation before the end of the war.

Two more ex-Soviet 12.2cm guns came within the orbit of the SS when the self-styled 'Russian National Liberation Army' (RONA) under its strongman Bronislav Kaminski who governed – in a rather medieval manner – the autonomous district of Lokot in Byelorussia, joined the Waffen-SS during the summer of 1944. The

Kaminski Brigade brought with it an artillery battery equipped with two Soviet 12.2cm gun/howitzers which had been acquired from the Red Army earlier in the war. Deployed to help SS security forces suppress the Warsaw Uprising on 1 August 1944, these guns contributed to the systematic levelling of large parts of the city. During these security operations, the *Kaminski* Brigade committed numerous heinous atrocities against captured Polish rebels. What incensed the German military command, however, was not these atrocities but several minor cases of indiscipline within Kaminski's force that affected German troops and civilians. On 19 August Kaminski was ordered out of Warsaw, and after a show trial quickly condemned to death and executed. The

BELOW: *A 15cm sFH heavy field howitzer on the Eastern Front in March 1942. The gun has been partially white-washed to provide concealment.*

ABOVE: *A heavy 17cm Kanone 18 about to be fired. In the Waffen-SS it saw service with only a select few corps artillery batteries.*

brigade was likewise withdrawn and subsequently disbanded – but only after it had completed its brutal annihilation of those Polish rebels still resisting the Germans! Waffen-SS artillery, therefore, not only contributed to that organisation's impressive battlefield performance, but also to the brutal suppression of resistance to German rule which the Nazi hierarchy required the SS to spearhead.

In the final analysis, Waffen-SS artillery was inferior when compared to its own armoured and anti-tank hardware. However, though there was resistance from the army concerning the Waffen-SS acquiring artillery, as mentioned above, this was not the main reason for the poor

SS artillery arsenal. The German artillery used in World War II was for the most part rather uninspired. Field and medium artillery, for example, were developed only by upgrading pre-war designs. This development proceeded along three lines: increasing the barrel length which, together with better propellant, gave greater muzzle velocity and accuracy; reducing the projectile weight to achieve the same effect; and utilising rocket assistance, though this was found to decrease overall accuracy. Thus artillery was a poor relation when compared to such equipment categories as anti-tank weapons, which developed apace under the stress of wartime. In addition to the lacklustre designs, Waffen-SS units also suffered from that perennial problem with regard to hardware during the war: there was never enough to go round. And captured stocks rarely made up the shortfall.

Abbott, Peter and Nigel Thomas, *Partisan Warfare 1941–45*, Osprey, London, 1983.

Ailsby, Christopher, *SS: Roll of Infamy*, Motorbooks, USA, 1997.

Bartov, Omer, *The Eastern Front, 1941–45, German Troops and the Barbarisation of Warfare*, St. Martin's Press, New York, 1988.

Bartov, Omer, *Hitler's Army: Soldiers, Nazis, and the Third Reich*, Oxford University Press, 1991.

Beinhauser, Eugen, Ed., *Artillerie im Osten*, W. Limpert, Berlin, 1944.

Bernhardt, Walter, *Die Deutsche Aufrüstung, 1934–1939*, Bernard & Graefe, Frankfurt a/M, 1969.

Butler, Rupert, *The Black Angels: The Story of the Waffen-SS*, Sheridan, London, 1978.

Chamberlain, Peter, and Ellis, Chris, *PzKpfW VI Tiger and Tiger II, (Profile AFV 48)*, Profile Publications, Windsor, Berks, 1972.

Chamberlain, Peter, and Doyle, Hilary, *The Panzerkampfwagen III and IV Series*, Iso-Galago, Bromley, 1989.

Cooper, Matthew, *The German Army 1939–45: Its Military & Political Failure*, Macdonald & Jane's, London, 1978.

Dunnigan, James F., Ed., *The Russian Front: Germany's War in the East 1941–1945*, Arms & Armour Press, London, 1978.

Edwards, Roger, *Panzer: A Revolution in Warfare, 1939–1945*, Arms & Armour Press, London, 1989.

Ellis, Chris & Doyle, Hilary, *Panzerkampfwagen*, Bellona, Kings Langley, Herts, 1976.

Emde, Joachim, *Die Nebelwerfer: Entwicklung u, Einsatz d. Werfertruppe im 2. Weltkrieg*, Podzun-Pallas, Friedberg, 1979.

English, John A., *The Canadian Army and the Normandy Campaign: a Study in the Failure of High Command*, Praeger, London, 1991.

Feist, Uwe & Nowarra, H. J. *The German Panzers from Mark I to the Mark V Panther*, Aero Publishers, Fallbrook, CA., 1966.

Feist, Uwe, *Deutsche Panzer 1917–1945*, Aero Publishers, Fallbrook, CA., 1978.

Fey, Willi, *Panzer in Brennpunckt der Fronten*, Munich, 1960.

Fürbringer, Herbert, *9SS-Panzer Division Hohenstaufen 1944: Normandie-Tarnopol-Arnhem*, Editions Heimdal, Paris, 1984.

Gander, Terry, *Small Arms, Artillery and Special Weapons of the Third Reich*, Macdonald & Jane's, London, 1978.

Gilbert, Adrian, *Waffen-SS: An Illustrated History*, Guild Publishing, London, 1989.

Grove, Eric, *World War II Tanks: The Axis Powers*, Orbis Publishing, London, 1971.

Guderian, Heinz, *Panzer Leader*, Futura, London, 1974.

Hastings, Max, *Das Reich*, Michael Joseph, London, 1981.

Haupt, Werner, *Das Ende im Westen 1945: Bildchronik vom Kampf in Westdeutschland*, Podzun, Dorheim, 1972.

Jars, Robert, *La Campagne de Pologne (Septembre 1939)*, Payot, Paris, 1949.

Jurado, Carlos C., *Foreign Volunteers of the Wehrmacht, 1941-45*, Osprey, London, 1983.

Keilig, Wolf, *Das Deutsche Heer, 1939–1945* (3 Vols.), Podzun, Bad Nauheim, 1956.

Keegan, John, *Waffen-SS: The Asphalt Soldiers*, Ballantine, London, 1970.

Kennedy, Robert M., *The German Campaign in Poland 1939*, OCMH, Washington D.C., 1956.

Koch, H, A, *Die Geschichte der Deutschen Flakartillerie*, Podzun, Bad Nauheim, 1954.

Landwehr, Richard, *The Lions of Flanders: Flemish Volunteers of the Waffen-SS*, Bibliophile Legion Books, Silver Spring, MD., 1983.

Lefevre, Eric, *Panzers in Normandy: Then and Now*, Battleline Books, London, 1984.

Lehmann, Rudolf & Tiemann, Ralph, *Die Leibstandarte*, (4 Vols.) Munin Verlag, Osnabrück, 1986.

Lucas, James, *Das Reich*, Arms & Armour, London, 1991.

Lucas, James, and Cooper, Matthew, *Hitler's Elite: Leibstandarte 1939-45*, Macdonald and Jane's, London, 1975.

Luther, Craig W. H., *Blood and Honour: the History of 12th SS Panzer Division 'Hitler Youth' 1943–1945*, Bender, San Jose, CA., 1987.

McLean, Donald B, *Illustrated Arsenal of the Third Reich*, Normount Technical Publications, Wickenburg, AR., 1973.

Mehner Kurt, ed, *Die Geheimen Tagesberichte der deutschen Wehrmachtführung im Zweiten Weltkrieg, 1939-1945*, Biblio, Osnabrück, 1984.

Meyer, Hubert, *Kriegsgeschichte der 12SS-Panzerdivision 'Hitlerjügend'*, Munin Verlag, Osnabrück, 1982.

Mollo, Andrew, *Uniforms of the SS*, (7 Vols.) Historical Research Unit, 1969–76.

Müller-Hillebrand, Buckhardt, *Das Heer, 1939–1945*, (3 Vols.) E. S. Mitler, Frankfurt am Main, 1954-69.

Niehorster, Leo, *WWII German Organizational Series*, L. Niehorster, Hannover, 1990.

Nowarra, Heinz. J., *German Tanks 1914-1968*, Arco, New York, 1968.

Pallud, Jean Paul, *Battle of the Bulge: Then and Now*, Battle of Britian Ltd, London, 1984.

Piekelkiewicz, Janusz, *Die 8.8cm Flak in Erdkampf-Einsatz*, Motorbuch, Stuttgart, 1988.

Quarrie, Bruce, *Hitler's Samurai: the Waffen-SS in Action*, Patrick Stephens, Wellingborough, 1986.

Quarrie, Bruce, *Hitler's Teutonic Knights: SS Panzers in Action*, Patrick Stephens, Wellingborough, 1986.

Reitlinger, Gerald, *The SS: Alibi of a Nation, 1922–45*, Heinemann, London, 1956.

Reynolds, Michael, *Steel Inferno: I SS Panzer Corps in Normandy*, Spellmount, Staplehurst, 1997.

Senger und Etterlin, Ferdinand M., *German Tanks of WWII: The Complete illustrated History of German Armoured Fighting Vehicles*, Stackpole Books, Harrisburg, PA, 1969.

Stein, George H., *The Waffen-SS: Hitler's Elite Guard at War 1939–45*, Cornell, New York, 1966.

Stern, Robert C., *SS Armor*, Squadron/Signal Books, Carrollton, TX., 1978.

Strassner, Peter, transl. David Johnston, *European Volunteers: 5 SS Panzer Division Wiking*, J. J. Ferorowicz, Winnipeg, Canada, 1988.

Sydnor, Charles, *Soldiers of Destruction: The SS Death's Head Division, 1933-1945*, Princeton University Press, Princeton, 1977.

Ullrich, Karl, *Wie ein fels im Meer: 3 SS-Panzerdivision Totenkopf in Bild*, (3 Vols.) Munin Verlag, Osnabrück, 1984–6.

Weidinger, Otto, *Division Das Reich*, (4 Vols.) Munin Verlag, Osnabrück, 1979.

Weingartner, James J., *Hitler's Guard: the Story of the Leibstandarte Adolf Hitler, 1933–45*, Feffer & Simons, London, 1974.

Williamson, Gordon, *SS: The Blood-Soaked Soil*, Brown Books, 1995.

Index